ACADEMY of
SCIENCES

Genetic Vulnerability of Major Crops

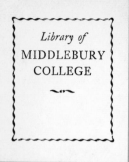

Genetic Vulnerability of Major Crops

Committee on
Genetic Vulnerability of Major Crops
Agricultural Board
Division of Biology and Agriculture
National Research Council

NATIONAL ACADEMY OF SCIENCES   Washington, D.C.   1972

NOTICE: The study reported herein was undertaken under the aegis of the National Research Council with the express approval of its Governing Board. Such approval indicated that the Board considered the problem to be of national significance, that elucidation of the problem required scientific or technical competence, and that the resources of NRC were particularly suitable to the conduct of the project. The institutional responsibilities of the NRC were then discharged in the following manner:

The members of the study committee were selected for their individual scholarly competence and judgment with due consideration for the balance and breadth of disciplines. Responsibility for all aspects of this report rests with the study committee, to whom sincere appreciation is expressed.

Although the reports of study committees are not submitted for approval to the Academy membership or to the Council, each is reviewed by procedures established and monitored by the Academy's Report Review Committee. Such reviews are intended to determine, *inter alia,* whether the major questions and relevant points of view have been addressed and whether the reported findings, conclusions, and recommendations arose from the available data and information. Distribution of the report is approved, by the President, only after satisfactory completion of this review process.

This study was supported by the Research Corporation of New York and by the U.S. Department of Agriculture.

ISBN 0-309-02030-1
Library of Congress Catalog Card Number 72-77533

Available from
Printing and Publishing Office
National Academy of Sciences
2101 Constitution Avenue, N.W.
Washington, D.C. 20418

Printed in the United States of America

ii

Preface

In 1970 an epidemic disease swept swiftly over the corn crop of the United States. A great agricultural resource of the country was threatened. In some sense, science and technology had been responsible.

The yield of corn dropped an estimated 50 percent or more in some southern states and 15 percent nationwide; that even greater losses might occur in subsequent years seemed possible. Memories of earlier plant disease epidemics made the corn blight even more alarming. For example, at the turn of the century, a chestnut blight epidemic had moved down the mountain spine of eastern North America, leaving it bereft of chestnuts. Innumerable "Chestnut Hills" remained on maps, but the chestnuts themselves were gone. Many still recalled the wheatless days of 1917, the great Bengal famine that killed thousands in India in 1943, and the Irish famine of the 1840's. Each had come about from the destruction of a staple food crop by plant disease.

No one really felt that the corn blight epidemic would cause famine in the United States; there were too many other high-carbohydrate crops for that. It did, however, prompt numerous questions: How serious? What about the next year? What happened? What caused the epidemic? Why was it not foreseen? Where did the technology go awry?

iii

In that it is a responsibility of the Agricultural Board to watch for perturbations in the nation's agriculture and to suggest means by which to reduce them, the Board established a committee to examine the blight epidemic. This group of plant breeders, plant pathologists, geneticists, entomologists, plant physiologists, economists, and men knowledgeable in the various major crops looked into the circumstances surrounding its occurrence and examined the more general issue of whether the genetics of other major crops is such that they are also vulnerable.

The report emerging from this study is divided into three sections. Part A examines the corn blight epidemic in detail: Technology is assessed; some historic epidemics and the phenomena of epidemics—the influence of the weather, the parasite, and the host—are examined; insect outbreaks are considered; and the economic implications of epidemics are assessed.

Part B deals with the major crops of the United States; that is, how important are they, how are they bred to improve yield and reduce costs, and to what extent may each be genetically vulnerable to epidemics?

Part C explores the challenges to science and to the nation that are posed by genetic vulnerability.

The report employs technical words where no alternative exists, but in each case defines them. Otherwise, scientific jargon was avoided in the hope that this study not be for scientists alone, but rather for citizens concerned about an important technical problem of the day.

We are grateful to the many scientific colleagues who have given graciously of their time and knowledge to this work and to Dr. Joyce C. Torio, the staff officer for the study.

COMMITTEE ON GENETIC VULNERABILITY OF MAJOR CROPS

James G. Horsfall, *Chairman*
George E. Brandow
William L. Brown
Peter R. Day
Warren H. Gabelman
John B. Hanson
Richard F. Holland*
Arthur L. Hooker
Peter R. Jennings
Virgil A. Johnson
Don C. Peters
Marcus M. Rhoades
George F. Sprague
Stanley G. Stephens
James Tammen
William J. Zaumeyer
*Resigned December 13, 1971.

Contents

Generalizations, Approaches, and Recommendations   1

I GENERAL CONSIDERATIONS OF EPIDEMICS

1	Anatomy of the Corn Leaf Blight Epidemic	5
2	Other Epidemics and Their Implications	17
3	Principles of Epidemics	23
4	The Pathology of Epidemics	43
5	Genetics and Epidemics	53
6	Dynamics of Insect Outbreaks	70
7	Economics of Epidemics	81

II VULNERABILITY OF INDIVIDUAL CROPS

8	Corn	97
9	Wheat	119
10	Sorghum and Pearl Millet	155
11	Rice	172
12	Potato, Sugar Beet, and Sweet Potato	190
13	Soybeans and other Edible Legumes	207
14	Vegetable Crops	253
15	Cotton	269

III THE CHALLENGES OF GENETIC VULNERABILITY

16	The Challenges of Genetic Vulnerability	285
	Committee on Genetic Vulnerability of Major Crops	305

Generalizations, Approaches, and Recommendations

- The nation's corn crop was struck by Southern corn leaf blight in 1970 because among other things a single source of cytoplasm had been utilized in developing a major portion of the corn hybrids.
- There suddenly appeared a new strain of a fungus pathogen well adapted to that Texas cytoplasm; weather being favorable, this fungus swept over the corn crop.
- Thanks to good corn weather in 1971 and to heroic efforts by seedsmen, scientists, and farmers, the epidemic that year was mild.
- The key lesson of 1970 is that genetic uniformity is the basis of vulnerability to epidemics.
- The major question the Committee on Genetic Vulnerability of Major Crops asked was, "How uniform genetically are other crops upon which the nation depends, and how vulnerable, therefore, are they to epidemics?"
- The answer is that most major crops are impressively uniform genetically and impressively vulnerable.
- This uniformity derives from powerful economic and legislative forces.
- The situation poses substantial challenges to scientists and to the nation.

1

• The scientist is challenged to be on the constant lookout for exotic pests and for parasite mutants that may attack our crops and to provide a back-up capability comprising diverse genes to be thrown into the breach as needed.

• The nation is challenged to provide a complete watchdog system, including

Overseas laboratories to monitor exotic pests and to test American varieties against them there—failing that, to establish such laboratories on offshore islands

A quarantine service at the borders (which it does)

A talent pool of scientists as the base for its efforts (which it does, in part)

A national monitoring committee to take the broad view

• The nation should provide (which it does, in part) the facilities for continuously maintaining gene pools, at home and abroad, as essential to the back-up system.

• The nation should (and does) provide insurance—by such means as crops in storage and crop insurance—against catastrophic losses.

PART I

GENERAL CONSIDERATIONS OF EPIDEMICS

CHAPTER 1

Anatomy of the Corn Leaf Blight Epidemic

Contents

A CASE FOR TECHNOLOGY ASSESSMENT	7
SEVERAL SCIENCES INVOLVED	7
THE STORY OF CORN BLIGHT	7
The Epidemic	8
Hybrid Corn	9
Detasseling	9
Pollen-Sterile Corn	10
The Restorer Gene	10
The Philippine Discovery	12
EXAMPLES FROM OTHER FIELDS	13
The Helminthosporium Blight of Oats	13
Drugs and Pesticides Lose Effectiveness	14
CONCLUSIONS	15

The epidemic of corn leaf blight in 1970 provides a good example for an anatomical study of genetic vulnerability. Although this epidemic has its own peculiarities, it illustrates the general principles of genetic vulnerability in all crops.

The story of hybrid corn is a proud one in American agriculture. It is well known to many Americans as one of the great contributions of science and technology to the feeding of an expanding population. Still further, the principle of hybrid vigor is not limited to use with corn. It can be generalized to such other crops as sorghum, sugar beet, carrots, and onions.

The corn crop fell victim to the epidemic because of a quirk in the technology that had redesigned the corn plants of America until, in one sense, they had become as alike as identical twins. Whatever made one plant susceptible made them all susceptible.

Uniformity is the key word—the plants were uniform in that special sense, and uniformity in a crop is an essential prerequisite to genetic vulnerability.

That a portion of hybrid corn technology went unexpectedly awry in 1970 puts it in the class with other unanticipated side effects of technology—a matter that has become of concern to citizens who

6

feel that scientists should ask the right questions about possible side effects. This has been called *technology assessment,* a term that has come into common use since about 1968. It is a subject with which scientists must deal. But it was not hybrid vigor as such that was responsible for the corn epidemic, it was a phase of the technology of using hybrid vigor.

A CASE FOR TECHNOLOGY ASSESSMENT

Not all the 1970 corn crop died—only part of it. Still, it was enough to be disturbing.

Against what benefits must this technological defect be appraised? The primary benefits of the technology of corn breeding are high. The corn yield per acre continues to rise and as of now shows no sign of abatement, except for the recent corn-blight episode. Yield has risen threefold since 1929. Much of this is due to breeding. Breeders can develop varieties of corn that are shorter and that can be planted closer to increase yield; they can make corn that uses fertilizer more efficiently; corn that can be planted earlier, cultivated more easily, and harvested mechanically.

Clearly, the 15 percent loss, more or less, in 1970 was not catastrophic, but warned that we must assess the technology of plant breeding across the board and act on that assessment.

SEVERAL SCIENCES INVOLVED

Disease and its epidemiology is the province of plant pathology, the science of the diseased plant, but it involves many allied sciences—genetics, climatology, ecology, entomology, biochemistry, and plant physiology. It is related also to the technologies of agronomy, plant breeding, seed production, fertilizer manufacture, and the like. The challenge is to coordinate all these skills as best they can be.

THE STORY OF CORN BLIGHT

The technology that resulted in the great epidemic of corn blight in 1970 passed through several stages over nearly six decades.

THE EPIDEMIC

The epidemic struck Florida first in the early spring of 1970 and spread northward in the wake of the greening wave that year. Some fields died almost completely, others only partially. When severely diseased fields were picked, clouds of spores boiled blackly above the machines. While the farmers worried, the plant pathologists assembled in concerned discussions; the corn futures market wavered uncertainly; and the national magazines enjoyed a field day reporting the troubles. All in all, it was an impressive event.

Two facts stood out: (1) Not all corn was infected—only those plants that descended from parents carrying what is known as Texas cytoplasm (Plate 1); and (2) the fungus responsible was *Helminthosporium maydis.*

Let us consider these in turn. Cytoplasm is the complex substance that fills every living cell, from bacteria to man. In it are to be found all the minute components of the cell—nucleus, chromosomes, mitochondria, oil globules, and so on. Its composition varies somewhat between species and varieties of living organisms, but to the extent that it determines susceptibility to blight, it is the same in every plant of the Texas strain of corn wherever it is grown, and in 1970 it was grown by nearly every corn farmer in America.

Helminthosporium maydis had been present in American corn fields ever since, and probably before, the days when Squanto showed the Pilgrims how to plant the crop. The very name *Helminthosporium* means that the reproductive bodies, called spores, resemble microscopic, segmented round worms, and *maydis* means that it occurs on maize. Before 1970 *Helminthosporium* sometimes blighted a few leaves and rotted a few ears, but it was not considered very important. It doubtless mutated from time to time, as all organisms do, and it probably produced more virulent strains from time to time. These strains, if they arose, tended to die out, however, because American corn was too variable to give the new strain a very good foothold.

The introduction of Texas cytoplasm changed all this. Corn was now nearly uniform throughout the country so far as its cytoplasm was concerned. While the technology of using T cytoplasm was being developed and spread across the nation, *H. maydis* continued to mutate. In due time, one of its mutant forms proved ideal for T cytoplasm and spread like wildfire across the cornfields.

The new mutant strain of *H. maydis* made a liability out of a technology that had been devised to improve the efficiency of the

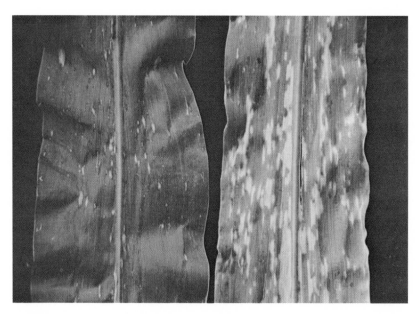

PLATE 1   Leaves of a corn hybrid with "Normal" cytoplasm (left) and the same hybrid with T male-sterile cytoplasm (right) showing contrast in reaction to infection by *Helminthosporium maydis*, Race T (Photo courtesy of A. J. Ullstrup, Purdue Univ.).

nation's agriculture. Losses in 1970 wiped out, for that year at least, some of the gain in efficiency in corn production acquired so laboriously during 50 years of research and its technological application. Let us now turn back the clock a half-century and watch the story unfold.

HYBRID CORN

In 1917 Donald F. Jones, a graduate student, discovered an aspect of hybrid corn whereby it could be produced commercially. Having first crossed an inbred line A with inbred line B and inbred line C with inbred line D, he then crossed the resulting AB with CD. Finally, he planted the ABCD seed and compared the yields with those of standard corn. They were up by about 25 percent—not by 2 or 3 percent, but by 25 percent! Now corn breeders could take their own locally adapted inbred strains and tailor the crop to fit any corn-growing region of the world.

Jones had established a principle, and a vast new technology got under way to put the principle to work commercially. It required some time, but by World War II hybrid corn had essentially driven the old corn varieties from American fields.

Looking back on the stages in the development of Jones's principle, one wonders about the technology assessment at each stage. Who assessed it? Who approved further development? How important was the role of science? How important that of economics?

DETASSELING

Jones produced hybrids by hand-pollinating his short rows of Connecticut corn. How then could this be done on thousands of miles of Corn Belt rows?

Corn is convenient in this respect. The plant has separate flowers for each sex—male in the tassel at the top of the plant, female in the ear about half-way down the stalk.

Seedsmen exploited this attribute. They planted six rows of the AB pair and then two rows of the CD pair, and so on, seed field after seed field.

The seedsmen developed a system employing thousands of high school students to pull out the tassels of the six rows of the AB corn, so that only CD pollen would fertilize the AB corn. This was what the economists call a labor-intensive technology and it was expensive.

Corn breeders assessed the technology and looked hard for a further technology to reduce the costs. Thus economics dictated further development.

POLLEN-STERILE CORN

One can imagine then, the significance to hybrid corn breeding when in 1931 Rhoades described a corn plant that had sterile tassels, in which the pollen was impotent! One could simply sow Rhoades' male-sterile plants in the AB rows, and they would be crossed automatically by pollen from the CD rows. No more detasseling! Cheaper seed!

Alas! All the progeny of this cross proved sterile. Sterility was inherited not through the genes, but through the cytoplasm of the female line, and became known as cytoplasmic male sterility. Had the farmer planted hybrid seed produced this way, he would have had only cobs. Since Rhoades's cytoplasmic male sterile could not be made useful and was troublesome to propagate, it was a technological blind alley and was lost.

Mangelsdorf and Rogers, however, soon discovered another cytoplasmic sterile mutation. Because they were Texans and this particular form of cytoplasm was found in a Texas corn variety, their strain was identified by a capital T—the Texas strain.

THE RESTORER GENE

Mangelsdorf, who had worked as a graduate student with Jones, gave some of the Texas cytoplasm to Jones, who decided in 1948 that he could use it with a restorer gene. He reasoned that since there are genes for nearly everything, there must be one that would restore cytoplasmic male sterility. He searched and found such a gene and put it into the CD pair, which carries the fertile pollen. When, therefore, this pollen fertilizes the male-sterile AB corn, it carries along the restorer gene, and fertility is restored to the sterile parent. The farmer can plant this seed and count on getting a crop.

Both hybrid corn companies and growers assessed this stage of the development and were pleased. No longer need the seedsmen hire countless hands to pull tassels from miles of corn rows, and the farmers could obtain seed at less cost. The farmer could no longer grow his own seed, of course (because the technology was too complex), but he was willing to give up this freedom. One guesses that consumers would have assessed the secondary effects as favorable, too, in that they could still be well and cheaply fed.

Let us look further at the technological assessment of the use of cytoplasmic male sterility in the production of hybrid seed corn. Since the outbreak of leaf blight, breeders have been criticized for having converted most of the commercial corn of the United States to this single source of cytoplasm. Our present knowledge does indeed indicate that this may have been a mistake. But this knowledge comes after the fact, and the extensive testing of comparable hybrids in normal and T cytoplasm, conducted prior to the release of T cytoplasm hybrids to the American farmer, is not generally known.

Breeders were, of course, aware that the incorporation of cytoplasmic male sterility into breeding stocks might well influence traits other than pollen production. To identify these effects if, indeed, they did occur, hundreds of thousands of dollars were spent testing hybrids that involved T and other male-sterile-inducing cytoplasms— tests that encompassed scores of environments over a period of many years. Hybrids using male-sterile cytoplasms were compared with their normal-cytoplasm counterparts for yield, for reaction to insect infestation, and for response to numerous disease organisms, including *Helminthosporium maydis.* The results were encouraging in that they showed that hybrids having the sterile cytoplasm were equal in performance to their genetic counterparts having normal cytoplasm— overall they were no better and no worse. As a result, the new method of producing hybrids, which eliminated the inconvenience of detasseling, was adopted and subsequently became widespread, both in the United States and abroad.

At the time it was made, this decision was justified. What was missing was an adequate knowledge of the potential genetic effect of cytoplasms. Past experience in plant breeding had generally shown that the cytoplasm carried by a specific genotype had little, if any, influence on important economic traits of crop species. In the course of breeding history, the differences between reciprocals were usually found to be minimal. In the light of this, it is hardly surprising that the corn breeders were not greatly concerned about the widespread use of a single cytoplasm. All this is now changed as a result of the observed susceptibility in 1969 and 1970 of T cytoplasm to *Helminthosporium maydis,* Race T. Breeders have now learned that in addition to nuclear genes, cytoplasms are an important aspect of diversity.

If society had been consulted, would they have approved this narrowing of the genetic base? We suspect they would have.

THE PHILIPPINE DISCOVERY

In 1962 two Philippine corn breeders, Mercado and Lantican, published a note in a Philippine journal, the essence of which was also published in the *Maize Genetics Cooperative News Letter.* In this note, they pointed out that in their country *Helminthosporium maydis* was highly virulent on several lines and hybrids carrying T cytoplasm. But neither in this paper nor in a second one published in 1965 did they warn of a possible epidemic. They had exposed the clue for a 1970 audience, but in 1962 it remained unseen by all. And why did the Filipinos not warn that the fungus could be damaging to all varieties having T cytoplasm? Perhaps because scientists are disciplined to avoid extrapolation. They probably reasoned, too, that they were working in a tropical environment not at all typical of the world's major corn lands.

A few vague clues turned up in America. In 1958 Jones stated the principle that genetic uniformity may lead to disaster in these words: "Genetically uniform pure line varieties are very productive and highly desirable when environmental conditions are favorable and the varieties are well protected from pests of all kinds. When these external factors are not favorable, the result can be disastrous . . . due to *some new virulent parasite.*" Jones did not, of course, warn against *Helminthosporium maydis,* but he did warn against uniformity.

Many have pointed out that more virulent races of parasites can arise in nature. One can now see that a 1965 paper by Nelson, on the genetic potentials of *Helminthosporium turcicum,* another parasite of corn, contained a clue. He warned that "more virulent biotypes can be expected to be preferentially selected and perpetuated," but he did not tie this to the Philippine discovery.

In 1958, even before the work of the Filipinos, Duvick reported that he could find no consistent differences between normal and cytoplasmically sterile forms of hybrids with regard to stalk breaking, leaf blight (caused by *H. maydis* or *H. turcicum*) or moisture percentage of grain. In 1965 Duvick published another paper, duly noting the Philippine work of a few years earlier and stating that to his knowledge no differences between T cytoplasm and normal cytoplasm had been reported or noted in the United States. "It may be," he wrote, "that the increased susceptibility in the Philippines is a secondary effect of reduced plant vigor, accentuated by the Philippine environment." The clue remained hidden.

As soon as the Philippine paper appeared, Hooker tested the possibility that T cytoplasm in his state could be especially susceptible to *H. maydis* and found no evidence of it. Because the results were negative, they were not reported in print until 1970. The situation had been assessed and any potential hazard associated with T cytoplasm appeared to be no greater than other hazards commonly accepted in the production of any crop. Only after he and associates had identified Race T as a new strain of *H. maydis* in the United States in the early months of 1970 did Hooker, in a paper written in the spring of 1970 and published in August state that "A majority of the acreage of America's most valuable crop is now uniformly susceptible and exposed to a pathogen capable of developing in epiphytotic proportions."

In any case, the stage was being set all during the 1960's. T cytoplasm was spread from Maine to Miami, from Mobile to Moline.

As in so many modern problems of environmentology, a study of the situation uncovers no villains. Instead we find a system where unseen forces carry well-meaning scientists toward a problem they had not intended or foreseen.

Have we learned enough about the complicated systems of agriculture to be no longer startled at the positive feedback when secondary effects of our well-intentioned technology arise? Can we learn to forecast these secondary effects, or at least to devise a system of negative feedback that will dampen rather than amplify our troubles?

The lesson to be learned seems to be that society in general, and scientists in particular, must realize that some changes of attitudes and emphasis will be required if the impact of technology going awry is to be minimized. Clearly, we must scan the horizon.

EXAMPLES FROM OTHER FIELDS

THE HELMINTHOSPORIUM BLIGHT OF OATS

The fate of the oat crop of the Midwest in the 1940's is a good example of the potential hazards of genetic uniformity. Oats had been plagued for a half-century prior to that with a disease called crown rust. In 1942 the plant breeders discovered a miracle oat called Victoria that stood up to all known races of the crown rust fungus and by 1946 derivatives of Victoria covered virtually the whole oat country. However, its lack of genetic diversity set the stage for an epidemic, since every oat plant carried the gene for crown rust resistance. In 1946 *Helminthosporium victoriae* arose in wild grasses and spread over the oat crop (Figure 1). Oats were not a major crop, but the blight was just as intense, if not more so, than the corn blight 25 years later.

DRUGS AND PESTICIDES LOSE EFFECTIVENESS

During the late 1940's and 1950's a drug or a pesticide would suddenly lose its effectiveness against the target organism. An antibiotic would suddenly fail to control a human disease; DDT would no longer control flies or a phosphate pesticide the red spiders.

This was precisely the situation for corn during 1970. Corn suddenly lost its effectiveness in warding off *Helminthosporium maydis.* The drug and pesticide people wrote books, held symposia, and developed three generalizations to explain how a chemical loses effectiveness: (1) It had been spread widely over the map; (2) it had been spread widely over time, i.e., it was a highly stable compound; and

FIGURE 1   Victoria blight (*Helminthosporium victoriae*) on oats. Left: a susceptible selection from Victoria carrying the crown rust resistance gene. Right: a resistant selection without that gene. (Photo courtesy A. Browning, Iowa State Univ.)

(3) a new strain of the target organism found a route around the road-block previously represented by the compound. They explained their results in one or both of two ways: (1) A pre-existing effective strain of the organism had been uncovered, or (2) the organism had mutated to form a new strain.

The same situation exists for corn if you assume that the chemicals in the Texas cytoplasm lost their effectiveness to ward off *H. maydis*. The conditions were the same—the Texas cytoplasm has been spread widely over the map; it had spread widely over time (i.e., over several years); and the roadblock had been easily bypassed. The fungus produced a toxin that broke down resistance of the T cytoplasm.

The plant pathologists even explained the rise of the new strain of *H. maydis* in the same way—either a pre-existing strain was uncovered or the fungus mutated. Most favored the latter explanation.

Farming is never free from hazards. These may arise from vagaries of the weather or from new diseases or insect pests. The possibility of an epidemic caused by *H. maydis* appeared no greater than that of other epidemics that could be caused by any one of a score of other pests.

The corn leaf blight epidemic is the first clear-cut case in plants where susceptibility to disease is inherited through the cytoplasm. Thus, breeders found themselves dealing not only with a new race of fungus, but with a new and unique form of disease susceptibility. As yet we do not have adequate biological methods to test differences in cytoplasm.

CONCLUSIONS

Crops become genetically vulnerable because of the uniformity society demands of the plant breeder. The market demands inexpensive food—the farmer who can produce it most cheaply captures the market. The market demands a uniform product—the farmer must produce it, and the plant breeder must produce the variety uniform in size, shape, maturity date, and the like. Uniformity of produce means uniformity in the genetics of the crop. This, in turn, means that a genetically uniform crop is highly likely to pick up any mutant strain of organism that chances to have the capacity to attack it.

Texas cytoplasm in hybrid corn is but one example of uniformity; this same principle will appear over and over again in the subsequent chapters.

By autumn it became clear that the 1971 season was not an epidemic year. The weather over most of the corn country was much less favorable to blight than was that of 1970. More important perhaps was the effectiveness of the return to former practices and the consequent availability of disease-resistant varieties.

CHAPTER 2

Other Epidemics and Their Implications

Contents

WERE THESE EPIDEMICS GENETICALLY BASED? 20

IMPACT 20

17

The leaf blight epidemic calls to mind other epidemics that have destroyed the crops of the world. It is useful to examine some of them. The consequences of past epidemics were sometimes tragic. Although our encounter with leaf blight in 1970 did not have severe consequences, it is easy to see how on another crop or in another country a future epidemic could be as bad or even worse.

The damage wrought by three epidemics is reflected in the Latin names of the causal organisms. The fungus responsible for the great Irish famine in the 1840's was named *Phytophthora* from the Greek words *phyton* (plant) and *phthora* (destruction). It destroyed both plants and people. A coffee rust epidemic in Ceylon in the 1870's was so devastating that the causal fungus was named *Hemileia vastatrix*. Similarly, the dread *Phylloxera* epidemic on French wine grapes stimulated the entomologist who identified the causal aphid to name it *vastatrix* also.

The history of man, in a way, mirrors the history of wheat rust epidemics. The Bible mentions them. In 700 B.C., the Romans created a god Robigus for the red rust. The French legislated against the barberry in 1660, a plant we know now to be an alternate host for the wheat rust fungus.

18

In 1916 the red rust of wheat destroyed two million bushels of wheat in the United States and another million in Canada, and the nation had two wheatless days per week in 1917. Rust epidemics spread across the wheat belt again in 1935 and in 1953.

The economic and cultural consequences of a rust disease are well illustrated by coffee rust. In 1870 Ceylon was the leading coffee nation of the world, exporting 100 million pounds annually, but by 1885 it was unable to export a single bag. The Oriental Bank failed, and the British became a nation of tea drinkers. Meanwhile, entrepreneurs had introduced coffee plants to South America, using vegetative cuttings from susceptible but disease-free plants. Here there was no rust, the industry flourished, and the peoples of North and South America became coffee addicts, the economy of several nations depending upon that fact. Except for one scare in Puerto Rico, the Western Hemisphere remained free of coffee rust until 1969, when the disease struck in epidemic form in Brazil.

The French wine industry was ravaged by three successive epidemics during the latter part of the last century—all three parasites came from America. In 1848 powdery mildew struck in force, and within five years had invaded most of the vineyards of Europe.

By the time the French got mildew under control with sulfur dust, there was another onslaught. The grape breeders, hearing that American grapes were immune to powdery mildew, imported thousands of these vines. With these vines traveled the root aphid that causes the *Phylloxera* disease, which proved worse than the powdery mildew.

In this instance, the French solved the problem by grafting their grapes onto American rootstocks. But in so doing, they imported millions more American plants and got for their pains another parasite, the downy mildew fungus, which devastated their vineyards for a third time.

The Dutch elm disease epidemic in the United States originated in the opposite direction. As elms died of this disease in Holland and France, American veneer factories bought the logs cheaply and in abundance. But these same logs scattered the disease-carrying beetles along the railroad rights-of-way out to the furniture factories in the Midwest, and the graceful elms that once lined the streets of America were lost in exchange for a few dollars' income for veneermen.

Tropical bananas, like French grapes, were hit by a succession of epidemics. The Panama disease, or wilt, moved into Panama from the Pacific shortly after the turn of the century and produced a catastrophic epidemic throughout the Caribbean, wiping out the tasty Gros

Michel variety and ruining many plantations. No sooner had this epidemic more or less subsided than the crop was attacked by another parasite imported from the Pacific Islands, the dreaded Sigatoka foliage disease.

The most recent serious epidemic disease of a food crop occurred in 1942 in Bengal. Here the fungus *Helminthosporium oryzae* devastated the rice crop and tens of thousands died the next year.

WERE THESE EPIDEMICS GENETICALLY BASED?

One cannot be certain, but the potato blight in Ireland was probably genetically based. The Irish grew mainly the susceptible variety Lumper, and since potato varieties are vegetatively propagated, the Lumper potatoes were all alike genetically.

There are no data on the coffee rust or the Bengal rice disease, but the *Phylloxera* disease of grapes could very well have been genetically based because the rootstocks were propagated vegetatively and could have been very similar.

The Dutch elm disease is probably not genetically based, but the phloem necrosis outbreak in midwestern elms 20 years ago almost surely was. It attacked almost exclusively the Moline elm, which was said to be a clone (i.e., vegetatively propagated).

The Panama disease of banana was surely genetically based because it attacked chiefly the Gros Michel clone.

The wheat rust epidemics of modern times are clearly genetically based in that as resistant varieties become available, farmers spread them over very wide areas. When the fungus mutates to a form that attacks the new variety, an epidemic ensues.

There has been much man-induced genetic vulnerability in the past. The Southern corn leaf blight epidemic is merely the most recent example.

IMPACT

The Bengal rice crop failed in 1942 under the combined effect of a typhoon and *Helminthosporium oryzae,* but the consequent high death rate occurred because India was then at war and had neither adequate food reserves nor means of transporting them to the needy. In much the same way the horrors of the Irish famine were compounded by the ignorance and inhumanity of the absentee landlords, the restrictive English Corn Laws, the lack of food distribution channels in the countryside, and the extreme severity of the winters of 1845 and 1846. In both instances, people were almost totally dependent on a single crop for their staple food.

Even before the coffee rust broke out in Ceylon, the Superintendent of the Royal Botanic Gardens there, Dr. H. H. K. Thwaites, said, "Ceylon will be ruined. Any country that gambles its future against the quick wealth that comes from a single crop is inviting bankruptcy." In other words, he was saying "avoid uniformity."

Had the United States depended on corn in the same way that Bengal depended on rice and Ireland on potatoes, the economic effect of the corn blight could have been severe indeed. Fortunately, the loss was much lower than it might have been. But many societies today in the undeveloped countries are still critically dependent on single crops. Millions of people in southern and eastern Asia depend on rice; millions more in Europe, northern Asia, and North America depend on wheat.

The protection of the human food supply requires constant effort. We are still striving to keep many diseases within tolerable limits. To control late blight, producers of modern potato varieties depend on chemical protectants and, to a slowly increasing extent, on resistance. Over the years our wheat crop reflects a succession of varieties that have different genes for stem rust resistance; the average life of each is little more than five years before it must be replaced with newer forms. The epidemic of Panama disease in banana has been contained for the time being by substituting more resistant varieties for Gros Michel. Epidemic diseases of forest trees present still more serious difficulties. To breed chestnuts for resistance to the blight disease (caused by *Endothia parasitica*) or white pine for resistance to blister rust (caused by *Cronartium ribicola*) takes many years. Other epidemics cannot be practically controlled in this way. We cannot expect to protect the suburban forests of the Northeast from epidemics of the gypsy moth (*Porthetria dispar*) by breeding for resistance—biological or chemical controls of the pest are more likely to succeed.

In the past, epidemics occurred because the host plants were uniformly susceptible to the parasite, the weather conditions favored their development, and the parasites spread. These encounters show clearly that crop monoculture and genetic uniformity invite epidemics. All that is needed is the arrival on the scene of a parasite that can take advantage of the vulnerability. If the crop is uniformly vulnerable, so much the better for the parasite. In this way, virus diseases have devastated sugar beet with yellows, peaches with yellows, potatoes with leaf roll and X and Y viruses, cocoa with swollen shoot, cloves with sudden death, sugarcane with mosaic, and rice with hoja blanca. Bacterial diseases have ravaged pears with fire blight and rice with bacterial blight. Nematodes have threatened potatoes while insects and fungi have played havoc with all these and many more.

To be unprepared is to maximize the impact. The more devastating epidemics are sudden. They catch us with no defense, no reserves, and no back-up potential. They hurt most when there are no alternatives. How great is the threat? How can it be minimized? How can the risks be made tolerable? Later chapters address these questions.

SUGGESTED READING

Carefoot, G. L., and Sprott, E. R. 1967. Famine on the Wind. Man's Battle against Plant Disease. Rand McNally, Chicago. 231 pp.

Large, E. C. 1950. The Advance of the Fungi. Johnathan Cape, London. 488 pp.

Van der Plank, J. E. 1963. Plant Diseases: Epidemics and Control. Academic Press, New York. 349 pp.

Woodham-Smith, C. B. 1963. The Great Hunger, Ireland 1845-49. Harper Row, New York. 510 p.

CHAPTER 3

Principles of Epidemics

Contents

THE DISEASE TRIANGLE	24
EPIDEMIOLOGY	27
The Timing Factor	28
The Spacing Factor	29
The Uniformity Factor	29
Time–Space Interactions: The Spread of Disease	30
Plasticity of the Parasite	30
The Exotic Parasite	31
QUANTIFICATION OF EPIDEMICS	32
The Model of Van der Plank	32
An Epidemic Simulator	33
Quantifying the Weather	36
Quantifying the Parasite (Inoculum)	39
PREDICTIVE SYSTEMS	40

23

We have seen that major epidemics have had and will have a devastating effect on crop production and, hence, on the stability of human societies. If we wish to prevent epidemics or at least to reduce the hazard of their occurrence, we need to know how they begin, what causes them to develop explosively, and how they run their course. This knowledge, rationally used, should allow for the timely application of controls. This, truly, is what the science of epidemiology is all about.

THE DISEASE TRIANGLE

To understand what causes epidemics it is necessary to understand the nature of plant disease. Considering only a single plant, there are always three factors that must come together in space and time in order for a disease to occur:

- The host—a susceptible host, of course.
- The parasite—a virulent, aggressive parasite that can readily gain entry into plant tissues, colonize them quickly, and reproduce (sporulate) abundantly.

24

• The environment—a favorable environment. Under the term environment a number of factors are included—rain, dew, relative humidity, temperature, wind, soil type, nutrition, light. For our purposes, however, environment can be virtually equated with weather.

Together, these factors form a "Disease Triangle"; when so depicted the nature of their coincidence and interaction, resulting in disease, is evident (Figure 1).

The fact that disease requires these three factors has been understood for less than 100 years. To ancient man, weather was the cause of plant disease and weather was thought to be the force that alone drove epidemics. Theophrastus, about 300 B.C., wrote that epidemics of plant disease developed in foggy river bottoms or during rainy periods.

Our knowledge of the cause of plant disease made few gains during the following 2100 years—the question was being hotly debated by the botanists of 1845 while the Irish slowly starved to death. The

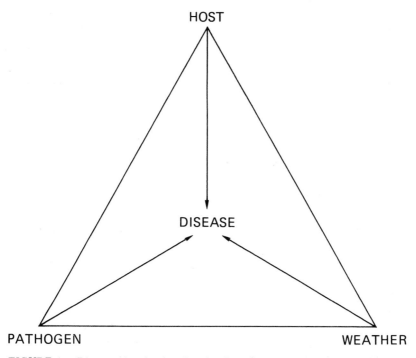

FIGURE 1   Disease triangle, showing the three factors that must coincide and interact if disease is to occur.

potato late blight epidemic years in Ireland were particularly foggy, rainy, and cold. Since an inordinately severe disease followed, surely the weather induced the disease. What could be more obvious? To be sure, the mycologists saw fungi on the rotted leaves of diseased plants, but they observed that the fungus always appeared after the disease. Because the fungus grew from putrified disease lesions, it was, *ipso facto,* produced by the disease, which was itself caused by cold, rainy weather. Theophrastus was right—or was he?

One striking exception occurred. Potatoes growing down-wind from a copper smelter in Wales escaped the epidemic, even though the weather was just as foggy, rainy, and cold as elsewhere. We now know that the copper released to the air and deposited on the potato leaves killed the fungus parasite, but at the time the incident did little to dispel the weather dogma. It was hard for the mycologists to believe that the fungus went on to produce new lesions and that without the fungus there would be no lesions. Those early scientists could clearly see the host and weather side of the triangle, but not the parasite.

In France almost 40 years earlier, Prevost had reported that a fungus could cause a plant disease—bunt of wheat. Although his proposal was rejected by the French Academy of Sciences as unsound, the potato disaster caused his evidence to be reconsidered. Within 20 years of the potato famine, the third leg of the triangle was firmly established for fungus parasites. Proof of the role of other parasites, bacteria, viruses, and nematodes, soon followed.

Man has never seemed to have trouble recognizing all three aspects of disease when the parasite was visible. The Bible clearly mentions the great locust (grasshopper) epidemics. The scribes knew full well that grasshopper epidemics occurred in dry weather, but they did not blame the plague on that alone. They could see the insect; they recognized that it caused the epidemic.

Of course, weather must be favorable. Had the weather in Ireland been dry in 1845 and 1846, there would have been no blight; and there was plenty of rain in the Corn Belt during 1970.

Understanding the disease triangle leads to some rather practical conclusions regarding control measures, for if three factors must coincide, it follows that disruption of any one or more of them must bring about control. Thus, if the environment can be altered so that it is unfavorable for disease, control is achieved. There are numerous applications of this principle in use today, including early or late planting dates and crop rotation and, for glasshouses, heating, cooling and protection from rain.

The use of chemicals—fungicides, nematocides, bactericides, eradicants, protectants, dips, drenches, fumigants, dusts, sprays, aerosols—serves to eradicate the parasite or to protect the plant, disrupting the coincidence of plant and parasite even in favorable weather.

With the studies of an Austrian monk, Gregor Mendel, in the late 1800's, the science of genetics was born. What we now term "Mendelian inheritance" came into use for the systematic breeding of plants. The significance of this event cannot be overstated, for now it became possible to breed plants resistant to disease and through this, in turn, to block the coincidence of plant and parasite.

Mendel's discoveries opened the door not only to an understanding of the genetics of plants, but also to an understanding of the genetics of plant parasites. Unfortunately, to this day the genetics of plant parasites, particularly as it relates to the behavior of populations of plant parasites in the plant–parasite interaction, remains little studied. It is unfortunate, because changes in the culture of crop plants cause significant changes in the population of plant parasites, and it is difficult to assess the genetic vulnerability of crop plants unless we can also assess the genetic potential and population behavior of plant parasites.

The biochemistry (and physiology) of the plant–parasite interaction is an area of study developed mainly over the past 20 years. Its significance in plant disease epidemics is not clearly understood, but the rational basis for chemical control of plant disease and, indeed, for breeding for disease resistance lies in the biochemistry of the plant–parasite interaction. If we can determine the nature of the interaction that allows for, or prevents, the rapid increase of disease over time, we will have at our disposal new, specific means for slowing the epidemic to a point where it is innocuous in terms of yield. The process of disease development follows a series of steps: arrival of parasite on plant surfaces, initial infection, an incubation period (the time when the parasite invades plant tissues), and sporulation (reproduction). Slowing or halting any of these steps can serve to so delay the onset of the epidemic, that serious loss does not occur.

EPIDEMIOLOGY

Epidemiology deals with the three factors of the triangle, to be sure; but rather than a single diseased plant, the science concerns populations of plants, populations of parasites, and "populations" of weather. Moreover, it deals with such populations in time and over space. Epidemiology deals with the rate of disease increase among plant populations.

THE TIMING FACTOR

A successful game of football depends upon timing. So does an epidemic. The events of an epidemic follow a sequence. If the timing is such that events occur in proper sequence, the epidemic moves. If the timing is off and sequential events are out of step, the epidemic is blocked.

The parasite also follows a sequence. Take the corn fungus, *Helminthosporium maydis.* It reproduces by spores that are formed on the disease lesion. The spores are blown by the wind to new corn plants. If the wind does not blow for a time, the spores cannot move, and the epidemic is slowed to that extent. When the spores arrive on the corn leaf, they must germinate and infect the healthy leaf. This process requires water. Without dew or rain at the proper time, the epidemic is delayed.

To obtain nutrient, the fungus must invade the tissues. A lesion develops, the tissue dies, and the fungus produces new spores, completing the cycle. If the weather is too hot or too cold, this stage in the sequence is slowed down. Under ideal circumstances the time from spore to spore takes about four days. Depending on when the disease started, this sequence may repeat itself as many as 10 times through the corn-growing season. If the weather is ideal for each stage in each sequence, the increase in disease over time is great—a major epidemic ensues. This took place during 1970.

The dynamics of an epidemic varies from one disease to the next. The increase of disease can be rapid, occurring within a few weeks—as in powdery mildew of glasshouse roses—or within a growing season—as with stem rust of wheat, late blight of potato or southern corn leaf blight. Increase may take place over a longer time, a few years—as with chestnut blight or white pine blister rust. Finally, it may occur over an extended period—as with Dutch elm disease or swollen shoot of cacao. No matter what the time, however, the major epidemic usually comes as a surprise to growers and scientists alike and usually causes major losses.

THE SPACING FACTOR

The greatest benefactor of an epidemic is crowding. If host plants grow close together, the fungus need not survive the hazards of travel. The new host is next door. For that reason isolation of an infected individual has been a classical method of dealing with epidemics of human diseases. "Pest houses" were used in the Middle Ages. Quarantine in the home is sometimes used even today. If an epidemic threatens, people are warned to stay out of crowds.

Corn, however, is crowded. Monoculture means merely the growing of a single crop species all together in one field. The cultivator is disadvantaged by distance between plants as much as the pathogen is. He wants them close together for easy tending and harvesting. Thus, his very system of agriculture encourages epidemics.

In recent years cultivators have increased the crowding of plants by using narrower rows and dwarf plants. Much of the "Green Revolution" is based on this concept—more plants per acre over more of the season to capture and store more of the sun's energy.

Crowding is an important factor in epidemics. The logistics of intense crop production inherently reduce the flexibility of crop production. In the case of southern corn leaf blight, changing the corn crop to something other than the highly susceptible Texas male-sterile hybrids requires time, and the epidemic hazard remains until that change can be made.

THE UNIFORMITY FACTOR

A factor related to crowding is uniformity in the population of plants. If there are but a few susceptible corn plants among the 18,000-20,000 plants on an acre, the parasite has the same distance problem as above. The susceptible individuals are likely to be widely scattered and an epidemic cannot occur. However, the demands of society for cheap food forces the modern cultivator to seek ever more uniformity. He must have the ears of corn at uniform height for his mechanical pickers. He wants the stalks to be of uniform height so as not to interfere with his weed-killing machinery. He wants his tomatoes to ripen simultaneously so that they may be mechanically harvested. If the genes for this uniformity happen also to make the crop more susceptible to a disease, an epidemic is in the making.

So it was with the corn blight epidemic of 1970. The corn uniformly had Texas cytoplasm. Only the right season was missing—1970 was the right season.

TIME-SPACE INTERACTIONS: THE SPREAD OF DISEASE

There is a critical relationship between time and space that bears directly on the rate of disease increase among uniform plant populations, although the interaction is not completely understood. Disease in corn, wheat, rice, or potatoes begins in discrete areas that we call foci. If the disease remains in foci, losses are directly proportional to the number of these and are usually unimportant. However, in an epidemic the disease spreads (over space) and usually spreads rapidly (in time) (Plate 2). The disease becomes general and losses are great.

How epidemics spread is a key question. It is related initially, of course, to the size of the parasite population and then, to the weather. Finally, and most important, it is related to characteristics of the crop itself. As the plants grow they form a canopy over the field and eventually produce their own weather—a microclimate. This in turn provides for a rapid increase in the amount of the parasite and for its spread. The result is an explosive outbreak of disease, an epidemic.

PLASTICITY OF THE PARASITE

The populations of most plant parasites are highly plastic, that is, they readily adapt to changes in the species and varieties of the crops. Thus, when resistant varieties are planted, the population of the parasite will shift. Often the previously occurring virulent form to which the crop was susceptible will "disappear" from the population. In time another member of the population capable of attacking the new resistant variety is likely to appear and to increase. This process, termed "selective pressure," or more recently "directional selection" by Van der Plank (1963), can result in a high population of the new virulent form of the parasite and precipitate an epidemic. Apparently, the stronger the resistance in the plant, the quicker the shift in the parasite population will be. The most typical illustration of this epidemic principle is in stem rust of wheat. Major epidemics of stem rust of wheat occurred in the United States in 1904, 1916, 1935, 1953, and 1954.

In 1926 the first strongly resistant popular common wheat variety, Ceres, was introduced and it was widely grown by 1930. In 1935 a hitherto unknown race of the stem rust parasite (Race 56) struck down the "resistant variety" and a major epidemic ensued. The wheat breeders soon developed varieties strongly resistant to Race 56, but in 1953 and again in 1954 another new race of the rust fungus appeared and another devastating epidemic occurred. Similar accounts could be given for other rust diseases. as well as for late blight of potato and, apparently, southern corn leaf blight.

THE EXOTIC PARASITE

When a parasite moves to another, distant area it often finds highly susceptible hosts with little or no resistance. In the history of human diseases a classic example was the introduction of smallpox to American Indians by the white settler. Classic examples of epidemics resulting from the introduction of exotic plant parasites are:

- Chestnut blight, introduced into the United States from the Orient on nursery stock about 1904.
- White pine blister rust, introduced from Europe, also on nursery stock.
- Dutch elm disease, brought from Europe on veneer logs.
- Blue mold of tobacco, introduced into the United States from Australia and then from the United States into Europe.
- Coffee rust, introduced into Ceylon and now into Brazil, where it threatens the entire coffee industry of South America.

The ever-increasing interstate, intercountry, and intercontinent shipment of plant materials (seed, bulbs, cuttings, fruit, etc.) raises the probability of new epidemics.

Thus, today, in the United States and certainly in other major crop producing countries of the world, the methods and technology of production are increasing the probability of major epidemics and serious crop losses. This is not to condemn the methods or technology; there is no other road to travel if we are to keep our complex society fed, clothed, and housed in a reasonably pleasant, aesthetic environment. We cannot return to the "good old days" and must, therefore, look to other alternatives.

One of the alternatives, a viable one, is to turn to the science of epidemic prediction. This can provide for an accurate assessment of the genetic vulnerability of crop plants, the genetic potential and population behavior of pathogens, and for a rational decision-making process—not only for determining the tactics of control, but also for directions in research. In short, we need to quantify the probability of the epidemic occurrence of plant disease. In this sense, studies of plant disease epidemics form the basis for strategic planning in the science.

QUANTIFICATION OF EPIDEMICS

Until recently, forecasting of plant disease epidemics has been qualitative rather than quantitative. The earlier forecasts were based entirely on the weather—for example, the forecast for late blight of potato, which is based upon an estimate of the behavior of the parasite in response to temperature and atmospheric moisture (rainfall, dew periods). On the basis of simple weather station meteorological data, growers are, in turn, instructed to apply protective sprays. Similar schemes have been developed for such other diseases as apple scab and downy mildew of grape. From data on soil temperature and rates of emergence of seedlings from soil, it has been possible to make successful forecasts of seedling blights of corn, wheat, and sugar beets.

The quantitative prediction of epidemics is relatively new. A sizable effort is being made to construct a mathematical model to predict the occurrence of Southern corn leaf blight. These models are difficult to construct because they involve the complex interactions of the disease triangle. The disease triangle is a convenient, schematic way of depicting the interactions between the parasite, the plant, and the environment; an epidemic, where populations of parasites interact with populations of plants and both interact with the environment, is more complex. One better comprehends the difficulty when one understands the science of meteorology and the prediction of weather phenomena. Long-range weather predictions are made with low-order probability; short-range predictions are made with a greater probability but may have little local significance. Because of the complexity of the interactions and because of the probability factor, particularly with long-range predictions, few have had confidence in the quantification of epidemics, and fewer still the courage to attempt it.

THE MODEL OF VAN DER PLANK

Although it is risky to assign "firsts" in science, it would appear that van der Plank (1960, 1963) was among the first to integrate the whole disease picture—genetics of plant populations, the behavior of populations of plant parasites, and the interaction of these with the environment—on a quantitative basis. No attempt is made here to treat van der Plank's work in detail. The student of epidemic quantification may wish to refer to his texts (1960, 1963, 1968). However, his mathematical approach is deterministic and does not involve chance. Recognizing that epidemics represent an increase of disease over time and that they follow a general pattern over time (Figure 2), van der Plank proposed that the rate of disease increase among plant populations can be accurately assessed by a series of deterministic expressions not unlike the expression of continuous compound interest with money, at least during the early period of the epidemic, when the increase is logarithmic. These expressions are very useful, not only in charting the progress of the epidemic, but in estimating the effectiveness of resistance, the impact of fungicidal sprays and the

PLATE 2 Spores of *H. maydis* carried by the wind through a gap in the trees produced this patch of disease in a southern cornfield in 1970 (Photo courtesy of L. Farrar, Ala.).

timing of their application, and the effectiveness of altering the environment. They integrate the interactions of the disease triangle; in measuring the rate of increase of the parasite's population, one also measures the condition of the host plants and the environment. Using these data one can determine whether, indeed, it is best to approach epidemic control through resistance (and what type of resistance), through the application of protective or eradicative fungicides, by changing the environment (planting dates, crop rotations, and the like), or by various combinations of these. For example, Merrill (1968, 1971) has plotted the infection rate of the Dutch elm disease for several cities in the United States, some using certain control measures and others using none. The data allowed him to compare the efficiency of controls and to estimate the projected numbers of trees lost with time (that is, how long it would be until the entire elm tree population of a city succumbed to the disease). These analyses provided a basis for decision making by the city fathers: whether control programs were delaying the epidemic sufficiently to justify the investment of tax monies or whether the delay of the epidemic was so brief that control was not economically justified.

In modifications of his basic mathematical expression, van der Plank attempts to deal with complex plant–parasite systems and proposes a series of formulae designed to probe the unknown and to stimulate investigations in the quantification of epidemics.

AN EPIDEMIC SIMULATOR

Van der Plank's approach makes the quantitative day-to-day prediction of an epidemic difficult. Such a prediction calls for a different tactic, epidemic simulation. Briefly, epidemic simulation attempts to quantify the hour-by-hour influence of the environment on the development of the epidemic. Detailed laboratory studies are conducted on the effect of the environment on the parasite and on the plant-parasite interaction.

Such studies include the influence of the environment on the infection process and on the sporulation process (for example, the influence of temperature, atmospheric moisture and light on spore germination, infection, incubation, symptom expression, sporulation, spread of spores, and the like). The data are then projected to the population in the field and an epidemic prediction made. Epidemic simulation can be accomplished only by using a computer to handle the mass of data involved and to calculate rapidly the probability of the occurrence of the event. Once the simulation of a particular dis-

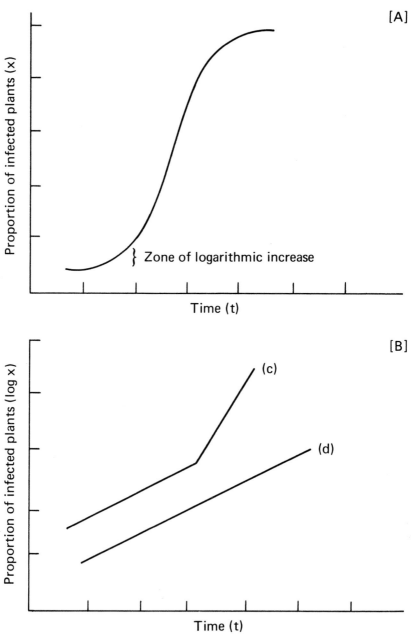

FIGURE 2   A: Arithmetic representation of the rate of disease in an epidemic. B: Logarithmic representation of the rate of disease increase in an epidemic — (c) changing rate of increase; (d) constant rate of increase.

ease is accomplished, however, it can be utilized for epidemic prediction of that disease wherever it may occur—assuming that the relevant weather data from the region concerned can readily be transmitted to the computer. Moreover, as with van der Plank's formulae, "war games" can be played with the simulator to answer strategic questions concerning specific aspects of control. Is it feasible to prevent the epidemic with fungicidal sprays? Will resistance that increases the incubation period or reduces the amount of sporulation effectively limit the development of the epidemic? Obviously, the answers to such questions will have a profound effect on the direction of research and upon the control measures applied.

Waggoner and Horsfall (1969) were the first to devise and publish an epidemic simulator, which they called *Epidem.* Using a model system involving tomato and *Alternaria* early blight, a fungus disease, they devised a computer program written in Fortran IV. They tested their simulator utilizing data from actual, earlier epidemics of the disease in Connecticut and found that *Epidem* predicted the occurrence of those epidemics with a high degree of accuracy. A few other simulators are now being devised: stripe rust of wheat is being so examined in the Netherlands; Waggoner and his colleagues in Connecticut have prepared a preliminary simulation of Southern corn leaf blight; other simulations are probably under study.

None of the epidemic simulations have yet been fully tested. They have, to be sure, been tried out using data from past epidemics reported in the literature; however, they have not been tested against the development of either naturally occurring or artificially produced epidemics. There is little doubt that the simulator will work, but the tests need to be made.

QUANTIFYING THE WEATHER

Epidemiology calls for the integration of each component of the disease reaction: the parasite and its genetic potential for attacking plants, the plant and its genetic vulnerability to attack, and the weather. It is the instrument for strategic planning, providing the theory and ultimately the means by which we can deal with future disease epidemics.

The literature of plant disease abounds with reports of the influence of climatic factors upon the growth and reproduction of the parasite and upon the disease reaction in a single or few plants. Fewer studies exist concerning the influence of the environment upon populations of plant parasites and upon large numbers of plants. Still fewer studies concern the influence of the environment on the rate of disease increase among plant populations. Van der Plank (1960, 1963) has treated the weather as it influences the rate of disease increase; Waggoner (1960) discusses the weather in relation to forecasting, as does Bourke (1970). Given a susceptible plant and a virulent, aggressive parasite the weather will obviously have a profound influence on the rate of disease increase and will be the determining factor in the development of the epidemic. If one examines the Frontispiece of *Epidem* (Waggoner and Horsfall, 1969) one can see at a glance exactly how the weather affects each stage in the development of *Alternaria* early blight (Figure 3).

If the weather is so important to the rate of disease increase and if it must be quantified in order to give accurate long- or short-range predictions of epidemics, how is it to be done?

In ascertaining the influence of the weather on the rate of disease increase we will have to deal with populations—this means dealing with the weather in relatively large-scale field experiments. To study weather and epidemic simulation it is necessary to make laboratory determinations of the effect of climatic factors upon the parasite and upon the plant-parasite interaction. To prove the simulation, relatively large-scale field experiments are then necessary. In both situations after the research is completed and the role of the weather "quantified," effective use depends on actual weather forecasts, and the accuracy of the epidemic prediction depends, in turn, on the accuracy of that forecast. Fortunately, experience with forecasting epidemics to date indicates that predictions will be successful. Let us take a closer look at the technology of quantifying the weather in the

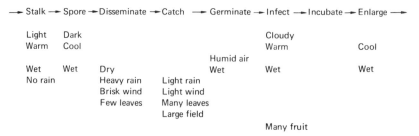

FIGURE 3 Diagrammatic illustration of the influence of weather on pathogenesis, from the epidemic simulator of Waggoner and Horsfall (1969).

field and laboratory so that we may better appreciate the magnitude (and cost!) of the job.

To "quantify" the weather in the field in order to determine its influence on the rate of increase in disease or to prove an epidemic simulation, rather sophisticated paraphernalia are required:

1. Sensors to determine the various climatic factors significant to the rate of increase of disease. These include light (quality, quantity and duration), atmospheric moisture (relative humidity), dew (duration) and rainfall (duration and amount), temperature (ambient, within the crop canopy and soil), and wind (velocity, direction and duration). These data should be recorded every hour during a 24-hr day and for as many days or weeks as necessary to follow the course of the epidemic.

2. Methods of detecting the initial inoculum (that is, the number of propagules) of the parasite and/or the rate of parasite increase. This, too, should be recorded each hour for 24 hr and for as many days or weeks as necessary to follow the course of the epidemic.

3. A data acquisition system at a point remote from the site of the epidemic that will accept the large volume of information transmitted from the climatic sensors and convert it into a form easily accepted and processed by the computer within reasonable time.

In the entire world there appear to be only two such systems available for studies on plant disease epidemics: one at the Agricultural University, Wageningen, The Netherlands; and one at Pennsylvania State University, University Park, Pennsylvania.

Climatic sensors and the data acquisition systems are readily available from commercial sources, albeit at a price, and the details of performance and mechanics readily obtainable from the sources cited. However, detecting the initial number of organisms and the rate of parasite increase (and spread) presents both theoretical and logistical problems that deserve further consideration.

Quantification of the climatic factors essential to the construction of epidemic simulators is best undertaken in the laboratory for several reasons, not the least of which is that such studies can be conducted year around. Thus, the simulator can be constructed in the winter and tested during the growing season. This is the case with the simulator for Southern corn leaf blight prepared by Waggoner and others, for example (See Felch and Barger, 1971).

Much has been published on the procedures and mechanics of such studies. Often the facilities necessary need not be sophisticated, but certain items are essential, given the basic laboratory and greenhouse facilities, including:

1. Apparatus for standardizing numbers of organisms (inoculum). The starting place for epidemic studies is the standardization of the inoculum needed for all subsequent experiments. Inoculum must be prepared and ready upon call in needed amounts for the studies. It must be prepared under defined conditions since its age and nutrient status is of key importance. Incubators with controlled, variable temperature and light regimes are essential.

2. Apparatus for studying atmospheric moisture. Atmospheric moisture (relative humidity, dew, rainfall and their magnitude and/or duration) is one of the most important climatic factors affecting epidemics. The time and duration of dew on the leaves, for example, is critical to the infection process, particularly germination and penetration. Relative humidity influences not only the formation of dew, but also the formation of spores, the quantity of spores formed, and the liberation of spores to the atmosphere. Rain affects the liberation of the spores of certain parasites and their spread over relatively short distances. It may also limit sporulation of certain parasites.

Studies of the effect of atmospheric moisture require special chambers with temperature controls, refrigerated walls to which heat from plant leaves can radiate so that dew may form, and temperature-controlled water baths to produce the desired relative humidities. It is useful, but not essential, to have lights in these chambers.

3. Apparatus for studying temperature, light, and wind. Temperature has a profound effect on the epidemic. It interacts with almost every other climatic factor (relative humidity, dew, light). It influences the growth and reproduction of the parasite, the growth and reproduction of the plant, and the plant-parasite interaction. There are optimum, minimum, and maximum temperatures for spore germination, for the penetration of the tissues of the plant, for the enlargement of lesions, for sporulation, and for spore viability.

The role of light, particularly light quantity and quality, in plant disease epidemics is not well understood. Light duration can be a determining factor in the reproduction of certain parasites. For example, *Alternaria solani,* the parasite used in the *Epidem* model, will not form spores in constant light at room temperature.

Air currents spread inoculum and, hence, are of great importance to parasites whose spores are airborne. In determining the spread of an epidemic from a local to a general area, the effect of air currents (wind, convection currents) must be known.

These climatic factors may best be quantified through the use of controlled environment chambers. Units are required that permit regulation of temperature, relative humidity and light (duration, quantity).

Moreover, the units must be so constructed that these factors can be varied with time. Thus, if it is desired to have a 20C, 14-hr day at 3000 ft-c with a relative humidity of 65 percent, followed by a 15C, 10-hr night with a relative humidity of 75 percent for a three-week period, the unit must automatically provide it once the controls are set.

Such factors as light quality, air currents, rain and dew can be studied by the use of smaller, specially designed units inserted into the larger unit.

Because the studies must determine, at the least, the minimum effect, the maximum effect, and the optimum effect and because several different phases of the disease must be studied (for example, spore germination, lesion enlargement, sporulation), a minimum of three each of the incubators, dew chambers, and controlled-environment chambers are needed. Otherwise, constructing a simulator will take months or years. Sophisticated computer systems that will accept, analyze, and store data are essential. Without them, quantitation of the weather for strategic and tactical epidemic prediction, cannot be accomplished.

QUANTIFYING THE PARASITE (INOCULUM)

The detection of initial inoculum can be important for quantifying the weather, whether in instituting artificial epidemics or in predicting natural ones. When plants are seeded with spores grown in the laboratory, the amount of initial inoculum can be directly measured. However, the inoculation process and the amount of inoculum used can affect the rate of disease increase and thus reduce the usefulness of the model.

Natural epidemics of diseases with high rates of increase start from but a small amount of inoculum, so small indeed that it is difficult to detect in the important, early stages of the epidemic. Furthermore, it is introduced over a period of time—so much on one day, so much the next, and so on. Thus, it may be difficult to determine initial inoculum and the proportion of infected plants early enough in the epidemic to determine the influence of the weather on the rate of increase at this critical time. At present there seems to be no way around this difficulty other than to count actual lesions on the site under study; a difficult, time-consuming, costly job.

To detect the rate of increase (and spread) is critically important to "quantifying" the role of the weather on the development of the epidemic. In general, it is accomplished in one or both of two ways: (1) Through spore-catching devices that can be placed at the epidemic site and through which the number of spores can be determined on an hourly basis to coincide with the recording of climatic data; or (2) by actual counts, or representative sample counts, of lesions on the plants. Both procedures are laborious; spore counting can be simplified by electronic enumerators, but to date nothing will substitute for manpower for lesion counts. The fact that the data obtained from either procedure must be transferred by hand to the computer card, further adds to the logistical problem. The critical issue is time: time to collect and count spores, time to count lesions. It may take weeks, even months, to quantify the effect of weather or to prove the simulator.

In the future this problem may be solved by remote sensing of the increase of disease. This is a rapidly growing science. The work of Manzer, Heller, Hoffer, Colwell and others indicates that remote detection from low-flying aircraft, using variously sensitive photographic film devices, is feasible. There is even some evidence that these techniques will detect the onset of foliar diseases, late blight of potato for example, before it can be detected visually on the ground. A major problem with remote detection is the degree of resolution of the detecting devices. We would like to "see" individual lesions on single leaves in the early phases of the epidemic. This would help determine initial inoculum as well as rates of disease increase.

PREDICTIVE SYSTEMS

The quantification of plant disease epidemics allows for strategic planning and for the rational development of tactics to avert or delay their effects. Epidemic quantification deserves consideration as a national (or international) goal. This, in turn, seems to call for an organized national (or international) effort, but what sort of effort?

As a specific example, consider the simulator. The quantification of a given epidemic, an epidemic simulator, requires a significant effort, but once it is completed it is applicable on a national (or worldwide) basis. Needed to make it so are: (1) A computer bank in which the simulator can be stored and retrieved at will; and (2) an interlocking of the computer bank with the national system of weather prediction. This, with proper computer programming, would allow, either automatically or upon call, for an "early warning system," on a local, regional, or national basis. Epidemic simulations as they now stand deal exclusively with predictions based upon weather, but there is no reason why genetic vulnerability of crop plants and genetic potential of plant parasites cannot also be incorporated. These factors should—indeed must—be so incorporated. In addition, it is important to understand that such systems can serve not only to predict epidemics, but to plan research. If, for example, the simulator predicts a major epidemic of corn in three to five years based upon genetic uniformity of the crop, upon the genetic potential of *Helminthosporium* for new, aggressive races, and upon a favorable long-range weather forecast, then a strategic decision can be made concerning research. Furthermore, the simulator can provide choices, on a probability basis, as to the direction of the research and, perhaps, even as to who should conduct it. Thus, with Southern corn leaf blight an immediate change in the genetic make-up of corn (from Texas cytoplasm to normal cytoplasm), accomplished largely by the industry, might be indicated.

The suggestion that a national predictive system be organized does not necessarily imply that an exclusive national laboratory be established. Nor does it imply that agriculturalists and scientists will no longer be surprised by major epidemics. A single laboratory, no matter how large, cannot hope to quantify all epidemics; much of the expertise lies in the states, and the necessary epidemic studies are best undertaken in the epidemic locale. A predictive system can collect, collate, and serve; and thus, can significantly reduce the likelihood of surprise. Bourke (1970) realistically examines predictive systems and his comments merit careful study.

Costs are and will be high. A quick estimate for minimal facilities and equipment would be in the range of $80,000–$100,000, exclusive of personnel, basic facility and computer costs, an amount that seems to call for regional and interregional coordination of research and for the establishment of research priorities. For example, high priority should be—indeed has been—given to quantifying the corn-*Helminthosporium* epidemic. Based upon assessments made elsewhere in this report priority should also be assigned to:

- Wheat — *Puccinia* (stem rust, stripe rust);
- Tobacco — *Peronospora* (downy mildew);
- Rice — *Pyricularia* (rice blast);

- Coffee — *Hemileia* (coffee rust);
- Potato — *Phytophthora* (late blight).

We have attempted to tell what is important to the occurrence of major plant disease epidemics. We have also attempted to tell how the newly emerging science of epidemiology is approaching solutions to the nation's potential epidemic problems. Because the science as described here is new, its significance to epidemic problem solving is not fully understood and resources for the necessary studies are limited. The Southern corn leaf blight emergency did stimulate new effort and additional allocation of resources, but only on an emergency basis and after the fact. Obviously, mere response to emergency is inadequate for the future.

REFERENCES

Bourke, P. M. A. 1970. Use of weather information in the prediction of plant disease epiphytotics. Annu. Rev. Phytopathol. 8:345–370.

Felch, R. E., and G. L. Barger. 1971. Epimay and Southern corn leaf blight. Weekly Weather and Crop Bull. 58(43):13–17.

Merrill, W. 1968. Effect of control programs on development of epidemics of Dutch elm disease. Phytopathology 58:1060.

Merrill, W. 1971. Elm disease control program question of aesthetics vs cost. Penna. Agric. Exp. Stn. Science for the Farmer 18:6–7.

Van der Plank, J. E. 1960. Analysis of epidemics, p. 229 to 289. *In* J. G. Horsfall and A. E. Dimond [Ed.], Plant pathology—an advanced treatise, Vol. 3. Academic Press, New York.

Van der Plank, J. E. 1963. Plant diseases: epidemics and control. Academic Press, New York. 349 pp.

Van der Plank, J. E. 1968. Disease resistance in plants. Academic Press, New York. 206 pp.

Waggoner, P. E. 1960. Forecasting epidemics, p. 291 to 312. *In* J. G. Horsfall and A. E. Dimond [ed.], Plant pathology—an advanced treatise, Vol. 3. Academic Press, New York.

Waggoner, P. E., and J. G. Horsfall. 1969. Epidem. A simulator of plant disease written for a computer. Conn. Agric. Exp. Stn. Bull. 698. 80 pp.

CHAPTER 4

The Pathology of Epidemics

Contents

INTRODUCTION 44
GENETICS AND THE PHYSIOLOGY OF DISEASE 45
NURTURE AND GROWTH OF THE PATHOGEN 47
Nutrition of the Pathogen 47
Pathogen Growth and Disease 49
HOST RESISTANCE 50
SUMMARY 52

43

INTRODUCTION

This chapter deals with the physiology of the disease itself. Here we must introduce a new term, *pathogen;* the discussion concerns the role of the pathogen in the induction of epidemics. A pathogen is an organism (usually a parasite) that initiates and drives a disease. We say *drive* because, for the disease to progress, the organism must stay with its host. If the organism leaves or dies, the host will recover. The damage will still be there, but the new growth, at least, will be healthy.

The word "pest" is an old one. "Pathogen," however, is a new word as words go. It comes from the Greek *pathos*—suffering—and *genesis*—to be born, to generate. A pathogen, then, is an agent (generally a living agent) that induces suffering.

Historically, if the causative organism is unknown or microscopic, the phenomenon has generally called an epidemic. If the organism is large enough to see with the naked eye, the occurrence is usually labeled an outbreak, only rarely an epidemic. We speak of an outbreak of corn borers but an epidemic of southern corn leaf blight. In line with this custom, agriculturalists have generally named the

44

little organisms pathogens and the big ones pests. (Since users of the English language are seldom consistent, a chemical to control either is called a pesticide.)

GENETICS AND THE PHYSIOLOGY OF DISEASE

Viruses, bacteria, fungi, and nematodes are infectious pathogens that use living plant tissue as the substrate for growth. If a plant had no mechanism for resistance, the pathogen would readily establish itself in the tissues and grow rapidly at their expense. The dead and dying host cells would map the progress of the disease as healthy plant tissues were damaged in the nurture of the pathogen. Shortly epidemics would be so severe that the host species would cease to exist—and so would the pathogen.

Evolution has eliminated the extreme of complete susceptibility, if indeed it ever existed. What is found is a very complex and delicate balance between the pathogens and their host plants. Populations of both survive in a dynamic struggle based on biological ploys and shifts to gain advantage. Measures and countermeasures result in uneven progress; surges of disease are followed by quiescent periods in which the parasite merely maintains itself.

The biology of the disease relationship has one very striking aspect: the pathogen is ordinarily restricted to a limited range of host species or varieties. By no means is every plant susceptible to every pathogen. Both host and parasite have genetic mechanisms for recognition and it is these mechanisms that determine the success of parasitism. A genetic change in the pathogen that makes it more successful, as in the case of *Helminthosporium maydis,* must be countered by a shift for resistance or evasion on the part of the host species if the latter is to survive. After eons of evolutionary change we see successful pathogens with efficient but narrowly based methods of parasitism, and hosts with a multitude of defenses left from innumerable silent conflicts for survival. The struggle is not over—we must maintain a genetic pool of competence for further change. A large amount of agricultural effort is devoted to identifying and selecting genetic variants for host resistance to pathogens, but what physiological property one should select is not always clear.

In the simplest terms, genes in both host and pathogen act by directing the types of enzymes synthesized, by which the type and severity of disease are controlled. The genes of host and pathogen are sets of instructions in a chemically written code, the deoxyribonucleic acid (DNA) of the nucleus. The coded information is transcribed during protein synthesis, and the enzymes thus formed catalyze the complex of biochemical reactions and syntheses that produce host and pathogen as we know them. Hence, the effect of genes on pathogenic events is only indirect. To understand the events themselves requires knowledge of the vital—or physiological—processes involved in parasitism. By what means does the pathogen invade the host and secure nourishment for colonizing the tissues? What physical and chemical defenses does the host possess to minimize or localize infection? We are specifically concerned here with the physiology of host and parasite at the point where the cells interact. If a host has genes for resistance, that resistance will be manifested in cell and tissue characteristics inimical to penetration, nutrition, and growth of the pathogen. The genetic problem of identifying and selecting genes for resistance is thus linked to the physiological problem of determining the mechanism of resistance. At present the metabolic pathways between genetic code and physiological expression are poorly understood.*

Molecular biology and molecular genetics, the general fields encompassing these studies, are still in their infancy despite spectacular advances. Membrane structure and cytoplasmic DNA are being intensively investigated, but there are great uncertainties about both. Mitochondria and plastids are small bodies in the cytoplasm that are concerned with respiration and photosynthesis respectively. They contain DNA, which appears to direct the synthesis of their own membranes; but nuclear DNA specifies most of the enzymes associated

---

*Resistance (or susceptibility, if you please) to Race T of the Southern corn leaf blight fungus (*Helminthosporium maydis*) is an exception to this general role of the genes. Here, the course of events in the host is governed mainly by the cytoplasm; the mechanism of this is far from understood. Resistance to Race T is conferred by a characteristic present in normal cytoplasm of various types of corn but absent or suppressed in Texas male-sterile Tms cytoplasm. Race T produces disease in the corn plant because it secretes a toxic substance that damages certain cell membranes, particularly those that bound the minute respiratory bodies (mitochondria) that float in the cytoplasm. The toxin is specific to the membranes in T cytoplasm; there must be a cytoplasmic factor that governs mitochondrial membrane properties so that Race T toxin reacts more readily with membranes from Tms cytoplasm than to those from other cytoplasms.

Understanding the physiological aspects of the disease rests on: (1) Determining the structure of the toxin and its mode of synthesis and secretion by the fungus; (2) discovering the membrane sites for toxin binding and the reasons for membrane failure; and (3) determining the difference between resistant and susceptible mitochondrial membranes. On the genetic side one must understand both the genetic control of toxin production by Race T and the genetic control of membrane synthesis in normal and Tms cytoplasm.

with these structures. In this connection, it is of interest that chromosomal genes modify the susceptibility of Tms cytoplasm to Southern corn leaf blight.

In summary, bridging the gap between genes and physiology requires intensive study of molecular genetics and cell biology. While these fields develop, a continued effort must be made to identify the physiological and biochemical attributes of resistance.

NURTURE AND GROWTH OF THE PATHOGEN

How does a pathogen grow and nourish itself at the expense of the host tissue? How does the host restrict or prevent this growth? What is the host's response to the stimulus of a pathogen? In the extensive literature on plant disease these questions are frequently interwoven, as might be expected in the study of a relationship. Since viruses, bacteria, fungi, parasitic plants, nematodes, and insects are disease agents, the subject becomes awesomely complex. For the purpose of this discussion, some generalized sorting out is needed. At risk of oversimplification, we will consider here only the more thoroughly studied of the fungal and bacterial infectious diseases. Facultative parasites and obligate parasites are considered separately. For insect-related diseases, see Chapter 6.

NUTRITION OF THE PATHOGEN

The nutrition of the pathogen at the expense of the host is difficult to study, and information on the subject is sparse. Studying nutrition *in vivo* presents special difficulties. Bacterial pathogens and many fungal pathogens are quite capable of growing on dead plant material or on synthetic media consisting minimally of a reduced carbon source (e.g., a sugar) and nutrient minerals. This is saprophytic growth, and the additional ability of these organisms to parasitize the host makes them facultative parasites or facultative saprophytes, depending on their dominant mode of nutrition. Growth under saprophytic conditions sometimes requires—or is promoted by—addition of such other organic materials as amino acids and vitamins. Presumably these organisms must gain at least the same substances from the host if they are to grow and reproduce. Hence, one possible avenue of resistance in a host would be a biological mechanism for denying the pathogen some critical metabolite. For example, plums resistant to *Rhodosticta quercina,* a fungus requiring myoinositol, are characterized by bark low in this metabolite. But such resistance based on nutrition is not generally regarded as of major consequence, except possibly in obligate parasites.

The nutrition of the pathogen outside the host is also important. To survive over adverse periods when the host is not available the pathogen must exist either as dormant spores and mycelia, or by growing on alternative substrates. Soil-inhabiting pathogens are in a favorable environment to survive as saprophytes on organic matter, and destruction of infected plant debris is a recognized prophylactic procedure. It is nearly impossible, however, to wholly eliminate infection sources from the environment and thus eradicate a disease. Destruction of host plants is the most effective means, but except for quarantine measures in a limited area, as for the citrus canker in Florida, it is not practical to destroy the crop in order to destroy the pathogen. Where alternate hosts are involved, as in wheat stem rust and white pine blister rust, their eradication has been attempted as a control measure.

Such obligate parasites as mildews and rusts do not normally grow outside the host, although several rusts can now be cultured on laboratory media. Clearly survival outside the host in nature must be as some form of dormant spore. If the obligate parasite kills its host, it cuts off its own food supply.

A characteristic of obligate parasites (and the nearly obligate facultative saprophytes) is the growth of specialized feeding structures called haustoria through the walls and into the host cells. In this way direct contact between the haustorium and the host protoplasm is established. The haustorium has a dense cytoplasm that contains numerous mitochondria (respiratory organelles), a condition indicative of intense metabolic activity. The close association of the two protoplasts provides the physical connection for nutrition, but the details of what is supplied, and how, are missing.

Not all pathogens feed so elaborately and delicately as this. Massive invasion of the host cells is common in the case of rot-producing pathogens. Seemingly, the invader secretes enzymes that degrade cell wall and protoplasm to obtain requisite nutrients. To some degree these secretion-digestion processes are common to all pathogens, otherwise they would not be able to penetrate between or into host cells. Plant cell walls are degraded by secreted enzymes that attack the pectins, hemicelluloses, and cellulose of which they are composed. In some plant diseases this process appears to be self-limiting, so that the host leaves and stem present a flecked, mottled, or streaked appearance due to a limited growth of the pathogen at each of a large number of infection sites. Field identification of a disease is often based on the character of the dead areas produced by the parasite. At the other end of the spectrum, it is possible to have infection and growth of the pathogen without any outward manifestation of disease. Indeed, there is one anomalous situation where *lack* of fungal infection and growth produces host plants that are nutritionally deficient and stunted: roots of certain species of higher plants must be infected with mycorrhizal fungi if they are to obtain adequate supplies of phosphate from the soil. This is an example of mutual benefit from the host-parasite relationship. The nutrition that the host supplies the fungus is repaid manyfold in better mineral nutrition of the host.

PATHOGEN GROWTH AND DISEASE

Inquiries into the relative susceptibility (or resistance) of the host plant to strains of the pathogen usually commence with a description of the penetration of the host by the pathogen, the infection and colonization of the host tissues, and the reproduction of the pathogen. Disease symptoms are described. From this point, inquiries branch out into such subjects as the causative agents for damage to host cells and tissues, and for host responses that confer resistance by suppressing pathogen growth. Virulent and avirulent strains of pathogen, as well as susceptible and resistant varieties of host plant, are used.

Penetration of root, leaf, or flower by pathogens starts with spore germination. Spores, like seeds, contain stored foodstuffs that permit germination and initial growth without drawing on external supplies. Spores will germinate if given a suitable temperature, adequate moisture, and oxygen. Field conditions thus become very important in regulating the level of infection. Disease can fluctuate between minor and epidemic proportions depending on the weather. Field resistance to disease through environmental control of spore dissemination and germination can be a major factor in host survival. There is an ecology of parasitism that warrants continued study.

On the plant surface the germ tube of a fungus will sometimes penetrate directly into epidermal cells, or alternatively enter the plant through natural pores (stomates and lenticels) or through wounds. Old leaves are more resistant to epidermal penetration, probably because of thicker cuticle and cell wall, or because inhibitory compounds have accumulated. Bacterial and viral infections often require wounds for penetration. Viruses can be inserted by sucking and chewing insects. Some substances secreted by host tissue tend to promote spore germination. Droplets of water surrounding spores on a leaf surface accumulate nutritive salts and metabolites that diffuse in from the cells below, and these can accelerate spore germination.

After penetration through the surface, the pathogen must grow through the tissue from the point of infection. Cells are penetrated by hyphae or haustoria. The pathogen secretes wall-macerating enzymes and various toxins that kill or damage the protoplasts of host cells. Many of the visible disease symptoms, as in the case of corn blight, have been ascribed to toxins. Toxins can diffuse ahead of the pathogen or be released in the xylem and moved through the plant in the transpiration stream. Toxins of several types are known or suspected. These vary from amino acid analogs to complex phenols and polypeptides. Their physiological action varies from a poorly described deterioration of the protoplast to the specific effect of making membranes leaky. Toxins also appear to affect the pathogen and are thought to limit pathogen growth in many diseases (hypersensitivity). Sometimes host cells participate in toxin formation by responding to injury by way of an aberrant metabolism that produces appropriate precursor substances.

One class of pathogens, the wilt fungi and bacteria, contribute to the injury of the host by producing materials that plug the water transporting cells of the xylem. The organisms themselves may contribute to this effect, but a major plugging factor seems to be gel-like materials produced as a result of wall degradation. In addition, as noted above, toxins can be released and carried to the leaves, where they act to alter membrane permeability. If the membranes become leaky, the leaf cells no longer retain their turgidity.

One of the more interesting types of damage concomitant with invasion and colonization is that in which the pathogen induces changes in the character of host cell growth. The pathogen either secretes a plant hormone or stimulates the plant to do so. Excess of hormone upsets the delicate balance required for precise control of growth. Galls and tumors can be formed, stems can elongate excessively, or alternatively, will thicken and twist. The aberrant growth lowers yields.

HOST RESISTANCE*

If the characteristics of resistance can be identified, there is hope of linking these genetically, thus placing plant breeding for resistance on

*For a further discussion of the genetics of resistance, see Chapter 5.

a firmer scientific basis. In all of biology and agriculture there is no more complex problem.

The resources that the host marshalls against invasion and colonization are varied. Nutritional deficiencies may slow the growth of the pathogen, and hence, of an epidemic. More important is the presence in the tissues of preformed materials toxic to the pathogen. The classic example here is the resistance of colored onion bulb scales to attack by the fungus *Colletotrichum circinans.* Dried scales of colored bulbs produce inhibitory phenolic compounds during their senescence and death; scales of white bulbs do not. Fungal attack, which starts by preliminary saprophytic growth on the dead scales, is prevented by the phenols—the correlation with color is fortuitous. In the case of root parasites, resistance through preformed chemicals can involve interactions between elements of the microflora. Root secretions sometimes favor the growth of beneficent or neutral organisms that antagonize the growth of pathogens, providing a second-order resistance.

The first line of host defenses is called upon in the initial stage of pathogen invasion. There are instances where roots, and sometimes leaves, appear to secrete organic compounds that can suppress germination and growth of a potential parasite. Phenols are frequently found to be the effective agents, not only against initial germination but also against growth subsequent to penetration. Mechanical means of resistance also exist in the form of thick cuticles and cell walls.

Mechanical impedance or antagonism to wall-degrading enzymes can play a part in resistance. Even such simple nutrient ions as calcium can be mobilized as antagonists. The toxins secreted by pathogens are thought in some instances to be antagonized or degraded by the host, although it is difficult to determine *in vivo* whether resistance is due to destruction of a pathogen toxin or to prevention of its formation. Wounds or lesions can be healed and essentially sealed off. There are ambiguities here: does healing occur only after pathogen growth is stopped, or does healing stop pathogen growth?

One of the more important resistance reactions of the host is the production of substances toxic to the pathogen in response to its presence. These host-toxins are called phytoalexins. Invasion and cell injury provide the stimulus, and the host cells are induced to synthesize fungitoxic compounds, largely complex phenols. There are many manifestations of changes in host metabolism in response to infection, of which the most common is increased respiration. Abnormal biosynthetic processes involving nucleic acid and protein synthesis have been found. Apparently, in some instances a genetic potential for biosynthesis of phytoalexins (and other compounds) is released by infection and the intensified host metabolism may reflect this. Nonpathogens or nonvirulent pathogens invoke the greater response— even wounding can be effective. Successful pathogens are poor inducers. However, sensitivity of the pathogen to the phytoalexin is also important, and some successful pathogens appear to tolerate higher levels of phytoalexins.

SUMMARY

The complexities of host-parasite relations are only touched upon here. The delicate balance between host and parasite, governed by manifold biological factors (e.g., nutrition) are all under genetic control. Our understanding of these factors is limited, as is our knowledge of the nature of gene action. Resistance is built into all crop plants, else they could not survive. We need to ascertain the quantitative and qualitative characteristics of resistance and the genetics of their inheritance. Without this knowledge we cannot economically conserve the important resistance features of genetic stocks.

SUGGESTED READING

R. K. S. Wood. Physiological plant pathology. Scientific Publications, Oxford. 1967.

CHAPTER 5

Genetics and Epidemics

Contents

GENETIC RESOURCES	54
Sources at Hand	55
Varieties from Abroad	55
Breeders' Stocks	55
Older Varieties	55
Wild Relatives	56
Induced Mutations	56
GENETICS OF RESISTANCE	57
Patterns of Resistance	58
One or a Few Genes; Many Genes; The Cytoplasm	
Genes and Environment	59
HOW PLANT BREEDERS WORK	59
HOST AND PARASITE INTERACTIONS	60
HOW RESISTANCE IS DEPLOYED	62
Different Aspects of Disease	62
Parasite Variation; Obligate Parasites and Facultative Saprophytes; Facultative Parasites	
Strategies	63
General Resistance; Specific Resistance; Multilines; General and Specific Resistance Combined	
Space and Time	64
Geographic Diversity; Resistance Genes Assigned to Regions; Variety Rotation	
Source and Dispersal of Inoculum	65
Source; Simple versus Compound Interest	
Crops and Cropping Practices	66
PESTICIDE RESISTANCE	68
GENERAL ASSESSMENT	68

53

To deal effectively with disease resistance, the research worker must clearly recognize two genetic systems: (a) That of the host; and (b) that of the pathogen or pest. The central problem in using disease resistance is to manipulate the genetic system of the host plant so that it can prevail over the genetic system of the parasite or pest under the environmental conditions where the crop is grown. In this chapter we are concerned primarily with the genetics of the host and how it can be manipulated by plant breeders to produce the type of crop plant that is resistant to diseases and insects while satisfying the grower and the consumer.

GENETIC RESOURCES

Resistant plants are essential to every program of breeding for disease and insect resistance. Such plants are to be found among the many types of cultivated and wild species of our crop plants at home and abroad. They are a reservoir of components from which the plant breeder can draw. They provide him with assorted constellations of genes, parts of chromosomes, whole chromosomes, sets of chromosomes and even cytoplasmic factors. If these are not sufficient, the breeder can generate additional genetic variation by treating his material with mutagenic agents.

Sources at Hand The first place to seek resistance is among the current native commercial varieties. If a resistant variety is found, it can very often be used immediately, as it is adapted to local conditions and acceptable to growers. In those instances where further breeding is needed, the objectives sought are more easily obtained than if a nonadapted source of resistance is used. As a rule, the agronomic, horticultural, and quality characteristics of adapted varieties are similar to those needed in the commercial variety. Undesirable characteristics are unlikely to be transferred with the resistance genes, and since the native commercial variety has already proved itself under local conditions, there should be little danger of accidentally introducing susceptibility to a rare disease or insect pest.

Varieties from Abroad Varieties from other countries or geographic areas have frequently been good sources of disease resistance. Because local strains of the parasite will not have had the opportunity to become virulent towards them, these varieties are, at least temporarily, highly resistant.

Improved varieties developed in one part of the world may give satisfactory yields in other parts of the world with similar climatic and soil conditions and day lengths. For example, corn hybrids adapted to the northern part of the United States Corn Belt are well adapted to some parts of Yugoslavia, and wheat varieties developed in Mexico have performed well in India and Pakistan. Hence, one country can sometimes take advantage of the plant breeding already done in another.

Breeders' Stocks An improved crop variety is the culmination of a series of steps achieved over years of effort. Some stocks are ordinarily left behind at each step. Although these stocks may be deficient in one or more essential characters, they can provide the source material for another cycle of breeding. In many plant breeding programs populations of source materials rich in genes for disease and insect resistance are developed and maintained. Although only a few varieties of a given crop may be grown commercially, the diversity of genetic types remaining in the breeding program can be substantial.

Older Varieties Before the development of intensive agriculture, many varieties and selections of a crop were grown. This is still true in certain less-developed areas of the world. Many have been collected and preserved in germ plasm banks and constitute an important reservoir of genes for disease and insect resistance.

Wild Relatives   Most major food crops in the United States have been introduced from a center of origin elsewhere. The wild relatives in and near the center of origin may have more resistance to disease than do their cultivated counterparts. In most instances only a small portion of the total genetic variability has been brought from the center of origin. The rest has come from local mutation. To recover some of the needed variability in a breeding program, one must cross the cultivated with the wild types, although this has its limitations in that the crosses are often sterile and the undesirable characters in the wild types may be difficult to eliminate. Nevertheless, wild relatives of tobacco, tomato, potato, sugarcane, wheat, oats, and other crops have contributed valuable genes disease-resistant to cultivated species.

Civilization poses an increasing hazard to these valuable wild types. "Improved" crop varieties are now being taken back to the centers of origin where they may replace local varieties and even occasionally the wild types themselves. There is real danger that this will greatly reduce the natural reserves of variation in wild types and primitive varieties that abound in regions where agriculture is not "advanced." In the current jargon, these are endangered species.

In addition, man is displacing the primitive forms of plants from the centers of origin with roads, altered drainage patterns, and other environmental perturbations. Thus, he alters the forces of evolution that have led to genetic diversity. Although plant collections and germ plasm banks are very useful in maintaining the genetic variability in these primitive forms, the genetic stocks still undergo erosion because not all collected material survives or is conserved. Protected areas, where relatives of major food crops can persist in their undomesticated state, are needed.

Induced Mutations   Certain chemical and physical agents induce mutations in plants and animals. Sometimes disease-resistant mutants can be found. These same agents may cause chromosome breaks that allow an alien segment of chromosome with desirable genes from a wild species to be incorporated into a chromosome of a cultivated species. In this way new genes for disease resistance may be made available to the plant breeder. Since mutations occur at random and most mutations are deleterious, the induction of mutations, however, has not proved to be a very reliable source of new germ plasm.

GENETICS OF RESISTANCE

Resistance to a parasite or insect can take various forms and is subject to several kinds of genetic control. Resistance is not an absolute quality but ranges from partial resistance to near immunity (Figure 1). It results from an incompatibility interaction between the metab-

FIGURE 1 Resistance (left) and susceptibility (right) to stalk rot in corn. (Photo courtesy A. L. Hooker, Univ. of Illinois.)

olism of the host and parasite that is governed by the genetic material of both. Genetic variation in the parasite is of great importance since a given host may be resistant to some forms of a parasite but completely susceptible to others. For many such cases each gene for resistance in the host may be matched by a corresponding gene for disease-producing attributes in the parasite.

PATTERNS OF RESISTANCE

The many studies of the inheritance of disease and insect resistance embrace a wide array of patterns but can be broadly classified into three categories: (a) One or a few genes; (b) many genes; or (c) the cytoplasm.

One or a Few Genes In some cases resistance may be determined by one or a few genes. Resistance that is due to a single dominant gene is very common and has been reported for a wide variety of plants against a wide variety of organisms including fungi, bacteria, viruses, mycoplasma, nematodes, and insects. Single-gene resistance has been employed extensively because it is easy to manipulate in a breeding program. Unfortunately, it has often proven to be a roadblock that is easy for the parasite to bypass.

Many Genes Resistance is often controlled by the joint action of many genes (polygenes), each with only a small effect. Crosses between resistant and susceptible parents show segregating populations that grade continuously from the level of resistance of one parent to the susceptibility of the other. Special statistical techniques and often special experimental designs are needed to distinguish polygenic resistance. Estimates of the nature of gene action, the relative heritability of the character, and the probable number of genes involved are possible. Polygenic resistance is more difficult to use in a breeding program than is resistance controlled by one or a few genes, but once selected and stabilized it tends to be long lasting as it is usually effective against a wide variety of parasite strains. Hence, it is good insurance against epidemics.

The Cytoplasm The importance of cytoplasmic inheritance is exemplified by the 1970 epidemic of Southern corn leaf blight. In this case corn with the Texas strain of cytoplasm, so widely used to produce hybrid seed by the male-sterile system, is susceptible, whereas all other tested cytoplasms are resistant to the pathogen involved.

Corn plants with Texas cytoplasm are also more susceptible to yellow leaf blight caused by *Phyllosticta maydis.*

In general, however, the genetic material affecting disease reaction in plants resides in chromosomal genes in the nucleus. Indeed, even in these two corn blights, such genes modify the disease reaction of Texas cytoplasm. Future studies may reveal more evidence for cytoplasmic inheritance of disease and insect resistance.

GENES AND ENVIRONMENT

Genes for disease resistance may respond differently to environmental conditions. For example, two genes may both give disease resistance at cool temperatures, but only one of the two give resistance at high temperature. Environmental effects may thus result in the occasional failure of some varieties to show their expected resistance. Such other environmental factors as light intensity and humidity are also known to modify disease reaction and may exert their effects on epidemics.

HOW PLANT BREEDERS WORK

Plant breeding is a science devoted to willfully and systematically altering the genetic materials of the plant to meet predetermined specifications for improvement. Historically breeding procedures were dictated by the mating system of the crop concerned. Crop plants may be cross-pollinated or self-pollinated. If the male and female flowers are borne separately, as in corn, the plants are cross-pollinated. If the two units occur in the same flower they are usually self-pollinated.

Breeders prefer to work with pure lines. With some plants this is difficult; with others it is easy. Plants that pollinate themselves, like wheat and tomatoes, show inbreeding of the closest type. Single plant selections provide the breeder with his pure lines.

Other plants, like corn, are cross-pollinated. Since the male and female flowers are on different parts of the plant, whatever pollen arrives first is effective and there are no natural pure lines. To obtain pure lines of corn the plant breeder artificially applies to the silk (pistils) pollen from the same plant. It may take several generations to achieve pure lines but once the lines are established, the breeder can proceed to "tailor make" the corn he desires by producing hybrids from the proper pure lines (inbreds).

Most breeding programs seek to combine the most useful genes by the most rapid and efficient methods possible. A number of methods are available, and there is little point in cataloging them here; they are adequately described in plant-breeding texts. Breeding programs lead rapidly to large-scale cultivation of the best products of selection and hence, to uniformity. While crosses between cultivated varieties and wild species promote genetic variability, they are commonly followed by methods that eliminate it, for example, inbreeding and selection for a predetermined plant type. Probably few breeding programs are designed to retain genetic variability within a variety. Indeed, this may not even be necessary or desirable if the range of varieties is sufficiently great. The danger is that for many crops the range is already narrow. Many breeders would argue that the deliberate retention of genetic variability within a variety would defeat their objective of uniformity. Perhaps for some crops, like wheat, a compromise can be reached through multilines. Even for crops like corn, where hybrid vigor requires the maintenance of different parental types, a narrow base of inbreds used in the production of hybrids can lead to dangerous uniformity. The results of cytoplasmic uniformity and the susceptibility of plants with Texas cytoplasm to Southern corn leaf blight are known only too well.

It has been argued that too few breeding programs emphasize new approaches. The new dwarf wheats and rice varieties from the International Rice Research Institute (IRRI) show that such programs can be very useful. The varieties represent a departure from older types to a new plant type with a wider environmental adaptation and a better response to fertilizer that in part explains the higher yields. They demonstrate the value of genetic conservation in the collections of varieties and species that provide the raw material for breeding. At the same time their successes threaten the survival of wild or primitive forms and their widespread planting leads to a genetic uniformity that should be avoided when resistance to disease and pests are concerned.

HOST AND PARASITE INTERACTIONS

For a long time it has been known that monoculture of genetically uniform crop plants on a large scale invites disaster from epidemics. Years of breeding for resistance to pests and diseases have shown a depressing pattern for many (but not all) crops: the successive introduction of resistant varieties followed by the development of new races of the pest or parasite capable of overcoming the resistance. Because of this, many support the view that all disease resistance is but temporary, that variability in most parasites permits the overcoming of resistance, and that breeding for disease resistance must be a continuing program. There are many examples, however, of resistance that has remained stable. This fact is often unrecognized. It is now becoming evident that parasite variability has been incompletely understood. Even regarding parasites that are comprised of differentially virulent races, there is considerable ground for optimism that disease resistance can be made more permanent, predictable, and effective in the future.

Genetic variation within the pest or parasite is extremely important to the success of disease control by means of host resistance. Disease resistance to a given parasite may take several forms: some may be specific to certain biotypes or strains of the parasite; others seem to function against all biotypes.

The forms of resistance that are effective against certain races (biotypes) of a parasite but not at all effective against other races are referred to here as *specific* resistance. (The phenomenon has also been labeled race specific resistance, differential resistance, or vertical resistance.) H. H. Flor (1956) in a series of classic experiments studied the genetic nature of the specific interactions of host and parasite in flax and flax rust. He proposed a gene-for-gene hypothesis that has subsequently been supported from studies in several other host-parasite systems. The gene-for-gene hypothesis simply states that for every gene that conditions resistance or susceptibility in the host plant, there is a corresponding gene that conditions virulence or avirulence in the parasite. There may be 25 or more sets of these corresponding genes. In flax rust, resistance occurs when any or all of the host genes for resistance are matched with parasite genes for avirulence. When resistance depends on only one gene in the host, a single gene mutation in the parasite is capable of rendering a resistant variety susceptible. Hence, specific resistance is unstable and likely to be lost as virulent biotypes of the parasite become common.

The forms of resistance that appear to be effective against all races of a parasite are referred to here as *general* resistance. (The terms generalized resistance, race nonspecific resistance, nonspecific resistance, uniform resistance, field resistance, and horizontal resistance have been used.) The nature of general resistance is not well understood and it may depend on several components. It is effective in disease control, appears to be of long duration, and can be used with little likelihood of failure.

HOW RESISTANCE IS DEPLOYED

Certain principles that govern the use of specific and general resistance in crops have emerged in recent years. Some are established; others are speculative; still others are novel and have not been tried. In many instances a combination of specific and general resistance or a combination of either kind of resistance with chemical protectants may be the most feasible course of action. The principles apply to parasites in general, whether they be fungi, bacteria, insects, or nematodes. Unfortunately, factors other than disease and insect resistance may determine the objectives of a plant-breeding program or the selection of varieties to be grown in commercial production.

DIFFERENT ASPECTS OF DISEASE

Plant disease control by host resistance works by keeping the parasite population at a low level so that disease does not develop to epidemic proportions. This is more successful for some diseases than for others and is affected by several factors.

Parasite Variation   Variation in parasites is a universal phenomenon. Such parasites as the potato late blight fungus, *Phytophthora infestans,* produce new races with a high frequency. Others, such as the *Fusarium* wilt fungi, have low rates of mutability. We can do little to prevent parasites from varying, but we can deploy host resistance genes so that the parasite gains no advantage from its natural variation.

Obligate Parasites and Facultative Saprophytes   The reproduction and survival of a parasite depends on its ability to compete for niches that provide food, water, air, and shelter. An obligate parasite, which cannot survive without a host, must find a niche in the host, whereas a facultative parasite may find a niche outside the host as well. Competition with other organisms is intense, and the ability of the parasite to survive in the absence of the host may be impaired by the many genes it must have to enable it to overcome host resistance. There is evidence that some resistance genes in the host exact a much greater penalty from the parasite in this respect than others. These have been called strong genes. Weak host genes exact little or no penalty, and the ability to overcome them has little effect on fitness of the parasite to survive.

Strong genes are important to a breeder. Unfortunately they are not common and they are not easy to recognize. When pathogenicity is not essential to survival, the competitive elimination of races pathogenic on hosts with strong resistance genes has been called stabilizing selection.

Facultative Parasites Breeding for resistance to facultative parasites, which have an active saprophytic stage, commonly in the soil, is relatively uncomplicated. Resistance to the *Fusarium* wilt diseases, caused by parasites in the water transport systems of many crops, has remained effective for 60 or more years and shows no indication of being in danger in spite of the appearance of new races of the parasite. These new races can be contained by crop rotation and by taking care not to spread the parasite.

STRATEGIES

There are many ways in which genes for resistance can be deployed in crops. General resistance or specific resistance may be used alone or in combination. Variations in space and time are possible. Varietal diversification may be practiced in local areas. There may be year to year rotation of varieties. Specific resistance genes may be assigned to different geographic regions based upon epidemiology.

General Resistance As a rule, no limits need be placed on the use of general resistance. As much as possible should be incorporated into each new variety. The limits would be those imposed by the level of resistance needed for adequate protection, by the genetics and expression of resistance, and by any incompatibility with other breeding objectives.

Specific Resistance The use of specific resistance alone puts the greatest selection pressure on the parasite. This selection is greatest when a single gene for specific resistance is widely used. The parasite responds with a gene allowing it to attack the new variety. When genes for specific resistance are used in units of two or three in each variety, the selection pressure on the parasite population favors those races that have the corresponding group of virulence genes. Such races, by chance, occur less frequently than do races with single-virulence genes. A still better protection against parasite variants is to develop complex varieties with many genes for specific resistance. This perhaps is most effective when such varieties are used initially rather than when they are developed and released by the breeder as a series of new varieties, each by the step by step addition of a new gene for resistance.

Multilines Multiline varieties are mechanical mixtures of lines in which the components are alike in genes for agronomic and quality characteristics but different in genes for specific resistance. Multiline varieties to control plant diseases were suggested first by Jensen (1952) and later in a different form by Borlaug (1959). The theory and methods have been developed more recently and a few multiline varieties released. When the multiline variety is released, it is expected that no one race of the parasite will attack all of the components. The resistant plants also inhibit the spread of the parasite between the susceptible components. This is somewhat analogous to using a mixture of antibiotics in human medicine. A super-race of parasite capable of attacking all of the components is not likely to develop.

The concept of multiline varieties can be extended to plant cytoplasms as well. The method allows for genetic and cytoplasmic diversity in resistance to diseases and pests without the loss of agronomic or horticultural uniformity. It is a conservative method of breeding in that the major expenditure of effort is on disease and pest control with little attention to improving yield, quality, and other characteristics. The method has considerable merit when excellent varieties already exist. It can also serve as a transition from varieties with a narrow genetic base for disease and pest resistance to varieties with broadly based and high levels of general resistance.

General and Specific Resistance Combined There are no genetic limitations to the use of general resistance in combination with specific resistance. In fact, each enhances the effectiveness of the other. In natural populations of wild plants both forms of resistance are present. When plant breeders select strongly for specific resistance, they frequently lose useful levels of general resistance. This is to be avoided.

SPACE AND TIME

Geographic Diversity Growing two plant varieties, each with a different form of resistance, in different fields, in a given geographical area, also reduces the risks of epidemics. The method has additional advantages when one variety, such as an early variety, is the source of parasite reproduction for a later variety. Under these conditions the spread of disease between fields, but not within fields, is reduced.

Resistance Genes Assigned to Regions   A strategy of regional distribution of genes for specific resistance has considerable merit in dealing with diseases where the parasite resides for a given period in one geographic region and then spreads along a definite course to end the season in another region. For example, in North America wheat rust fungi overwinter on winter wheat in the southern part of the United States and northern part of Mexico. From here the parasite spreads progressively northward until the spring wheats of Canada are infected. If different genes for resistance were allocated to specific regions along this pathway, the course of disease development would be slowed and might not even reach the spring wheats of the North. The system is based on the fact that the inoculum reaching each new region would be comprised mostly of avirulent races, since these would be the races favored for survival in the preceding region. To become effective such a program would necessitate careful planning and the full cooperation of all research workers and farmers involved.

Variety Rotation   The suggestion has been made that different varieties be planted in rotation as a means of better utilizing genes for resistance in a given area. A particular resistant variety would be planted, say, one year out of four. It would not be grown at all for the remaining three years. Strains of pathogens, even though they adapt to last year's variety, would be exposed to an entirely new variety each year and hence would gain no advantage from this variation. It is doubtful if variety rotation, as just described, will ever be practical. If stabilizing selection would result in the elimination of virulence genes in the parasite when they are not needed, however, formerly obsolete genes for specific resistance would repeatedly become useful again for plant breeding in a later succession of varieties.

SOURCE AND DISPERSAL OF INOCULUM

Sources of inoculum, means of dispersal, and type of epidemic are also important considerations in the effectiveness of host resistance as a means of control.

Source   Van der Plank in *Disease Resistance in Plants* (1968) makes the very important generalization that a minimum of two host genotypes are needed for specific resistance to exist. It follows then that specific resistance will not be effective if the parasite survives continually on plants of the same genotype or apart from these plants only as a resting stage. If an epidemic starts with the parasites coming from an outside source, and if the parasite persists there from one crop season to the next, specific resistance can be used effectively. For example, in stripe rust of wheat in the Pacific Northwest, the fungus each year comes from wild grasses (where it also persists) to the wheat crop. The wheat is on the receiving end of the epidemic, and the resistance genes in wheat exert little or no selection pressure on the residual parasite population in the wild grasses. Other examples are nematodes and certain viruses that persist on weeds or other crop plants and spread from these plants to the commercial crop.

Simple versus Compound Interest   Populations of some parasites increase at a rate comparable to simple interest on money; others increase at a rate comparable to compound interest. Simple-interest diseases, of which wilt diseases and the cereal smuts are typical, do not continue to spread from diseased to healthy plants during the course of a growing season. Compound-interest diseases, such as corn blight and cereal rusts, spread repeatedly. Specific resistance can be effective against the simple-interest disease. By contrast, if a new race occurs in the parasite population as it did with Race T of *Helminthosporium maydis,* it can spread across vast crop regions within a few months. Here, specific resistance is not likely to succeed. In general, foliage parasites with an efficient aerial means of spread are less easily controlled by specific resistance than are soil-borne parasites where new races, if they appear, spread more slowly.

CROPS AND CROPPING PRACTICES

Parasites of annual plants must persist apart from the host plant for a portion of the year. Compared to saprophytes parasites tend to be at a serious disadvantage at this time; hence the population of the parasite is reduced. This advantage is missing in perennial crops, and persistent resistance is more difficult to achieve than it is in an annual crop. Perennial crops are also more difficult to breed; hence the short-lasting specific forms of resistance should not be used.

Specific resistance delays the onset of an epidemic if races virulent to the resistant host are absent or rare in the parasite population. When the parasite is carried in vegetative propagating material, this advantage is lost since the host is infected at the start of the season. For compound-interest diseases a few infected plants can provide a large enough number of parasites for an epidemic. Hence, specific resistance is unlikely to be effective against compound-interest diseases that are carried on the seed or other propagating material of the host. Here, general resistance has a clear advantage.

The potential for epidemic disease development is greater in self- than in cross-pollinated crops, because the latter are genetically more variable than self-pollinated crops. With the development of single-cross hybrids, much of the advantage of this genetic variability has been lost from corn, ordinarily a cross-pollinated crop. The same is true for other plants. The breeding of cross-pollinated crops is easier to accomplish than is the breeding of self-pollinated crops when the character under selection is polygenic in inheritance. Hence, general resistance, often polygenic in inheritance, may be rare in self-pollinated crops that have undergone varietal improvement by man for many years.

Crop size also influences the probability that specific resistance will succeed. In general, when a crop is grown over a wide geographic area, the organisms causing foliage diseases tend to reproduce and come from within the crop itself. If, on the other hand, the crop occupies only a small area within regions predominately devoted to other crop species, the probability that the initial parasites would reproduce and come from outside the crop is increased. In these situations, stable resistance is more easily achieved even though specific forms of resistance are used.

Plant parasites frequently persist from one season to the next in the soil or in plant refuse. When nonhost plants are grown for one or more years the population of parasites is reduced and this aids in resistance. An extreme example of this is in sugar beets, where tolerance (ability to yield even though infected) to nematodes is the only form of resistance known. Tolerance is effective as a control only if used along with crop rotation. Similarly any crop management practice that reduces the population density of parasites between growing seasons enhances the expression of resistance. Thus, chemical sprays may aid resistance if they are effective in reducing the population of parasites.

PESTICIDE RESISTANCE

Many epidemics are held in check by fungicides. These are chemicals that interfere with the pathogen by preventing its entry into the host, restricting its growth, or reducing its ability to form spores. Insecticides are also widely used to combat insect pests. Pesticides, like specific resistance genes, introduce problems. Not only are we increasingly concerned about their effects on the environment, but they do not control pests that develop resistance, a phenomenon that is known in some 280 different insects. While resistance to fungicides is not nearly so common, it is a cause for concern. The genetic and biochemical basis for pesticide resistance has been worked out in a variety of target organisms. Insects and fungi have a remarkable ability to produce mutants that can detoxify or transform toxic chemicals. It seems that the more widely these materials are used the greater the selection pressure for resistant forms will be.

Among fungi we have already encountered citrus storage molds resistant to biphenyl; stinking smut, caused by *Tilletia foetida,* resistant to hexachlorobenzene; apple scab resistant to dodine; and seedling blight of oats, caused by *Pyrenophora avenae,* resistant to organomercurial seed dressings.

Pesticides will continue to play an important role in protecting our crops from epidemics but, like a single gene for resistance, we cannot often afford to rely on a single agent. Nor should we rely solely on pesticides as we have tended to do for insect control. In many crops the two can be integrated until protection may be largely achieved through innate resistance. Progress in this direction is reviewed in the chapters on individual crops.

GENERAL ASSESSMENT

Breeding for resistance to pests and diseases has been practiced for over 75 years. Few anticipated the nature of the problems encountered. Certainly the capacity of parasites to adapt to changing host varieties and certain forms of resistance was not appreciated. It is now clear that host resistance genes need to be classified as to specific resistance or general resistance. This is difficult to do. It is also evident that genes for specific resistance should only be used as one component of a broadly based system of resistance if maximum value is to be obtained.

Robinson (1971) has discussed the principles to be considered in the use of specific resistance and has produced a set of useful rules or considerations. These define circumstances under which specific resistance is likely to fail and to succeed. These rules and other considerations that might be added will indicate for a given plant disease the advisability of using specific resistance.

Man cannot prevent plant parasites from varying. He can, however, utilize the resistance genes to their maximum value so that the development of epidemics is minimized. Although diseases may be similar, no two have exactly the same epidemiology. Each disease needs to be considered separately when the deployment of resistance genes is considered. Man's skill in doing this will rest upon his understanding of relevant biological concepts and his ability to put this knowledge into practice.

REFERENCES AND SUGGESTED READINGS

Borlaug, N. E. 1959. The use of multilineal or composite varieties to control airborne epidemic diseases of self-pollinated crop plants. First Int. Wheat Genet. Symp. (Winnipeg), 1958. Proc., 12–26.

Browning, J. A., and K. J. Frey. 1969. Multiline cultivars as a means of disease control. Annu. Rev. Phytopathol. 7:355–382.

Day, P. R. 1966. Recent developments in the genetics of the host-parasite system. Annu. Rev. Phytopathol. 4:245–268.

Flor, H. H. 1956. The complementary genic systems in flax and flax rust. Adv. Genet. 8:29–54.

Frankel, O. H., E. Bennett, R. D. Brock, A. H. Bunting, J. R. Harlan, E. Schreiner [ed.]. 1970. Genetic Resources in Plants. Davis, Philadelphia. 554 pp.

Georgopoulos, S. G., and C. Zaracovitis. 1967. Tolerance of fungi to organic fungicides. Annu. Rev. Phytopathol. 5:109–130.

Hooker, A. L. 1967. The genetics and expression of resistance in plants to rusts of the genus *Puccinia*. Annu. Rev. Phytopathol. 5:163–182.

Hooker, A. L., and K. M. S. Saxena. 1971. Genetics of disease resistance in plants. Annu. Rev. Genet. 5:407–424.

Jensen, N. F. 1952. Intra-varietal diversification in oat breeding. Agron. J. 44:30–34.

Leppik, E. E. 1970. Gene centers of plants as sources of disease resistance. Annu. Rev. Phytopathol. 8:323–344.

Painter, R. H. 1951. Insect Resistance in Crop Plants. Macmillan, New York. 520 pp.

Robinson, R. A. 1971. Vertical resistance. Rev. Plant Pathol. 50:233–239.

Watson, I. A. 1970. Changes in virulence and population shifts in plant pathogens. Annu. Rev. Phytopathol. 8:209–230.

Williams, W. 1964. Genetical Principles and Plant Breeding. Davis, Philadelphia. 504 pp.

Van der Plank, J. E. 1968. Disease Resistance in Plants. Academic Press, New York. 206 pp.

CHAPTER 6

Dynamics of Insect Outbreaks

Contents

FACTORS REGULATING INSECT POPULATIONS   71
   Autoregulation   72
   Density   73
   Vigor   73
   Weather   73
HOW INSECTS FIND THEIR HOSTS   74
COMPUTER SIMULATION   76
SIGNIFICANCE OF HOST UNIFORMITY   77
INSECTS AS DISEASE VECTORS   78
INSECT QUARANTINES AND SURVEILLANCE   79

70

Insect outbreaks have troubled man throughout recorded history: witness the plagues of gnats, flies, and locusts recorded in the Book of Exodus and the mass devastation described by the Prophet Joel. These writers and many that followed concerned themselves with "why" outbreaks occurred. Scientists prefer to ask "how"—that is, they pose quantitative questions about insect outbreaks. This discussion will be limited mainly to the insect–host plant relationship, but will touch briefly on some other factors that influence insect populations. Specific insect–crop vulnerabilities will be considered in appropriate chapters devoted to individual crops.

FACTORS REGULATING INSECT POPULATIONS

In characterizing insect populations and insect outbreaks, a very useful concept has to do with the numbers of insects in relation to the unit of crop plant attacked. The relationship between insect numbers and crop damage is not a simple linear function. Very low populations may even be beneficial to the crop in stimulating tillering or other aspects of productivity. Once adjustment is made for these low levels,

71

there is usually a range of populations in which the ratio of insect density to crop damage approximates a straight line. Within this range predictions can be made relating numbers of insects to crop potential, which has been defined as the yield that can be produced with the maximum use of available technology. Certainly such other factors as growing conditions, stage of plant development, and plant spacing must be considered in addition to the number of insects.

General equilibrium position (GEP) is the phrase used to describe the density of an insect population over a period of time in the absence of permanent environment change. It is assumed that the GEP is arrived at after a long series of fluctuations in numbers. To be sure, fluctuations continue, but they are more restricted about the mean population level, which may be above or below the "economic injury level," which is in turn defined as the lowest population density that will cause economic damage. We are not very concerned for those insect species in which the GEP is below the economic injury level. Although many species fall in this low category, the few in which the GEP is well above the economic injury level are those that cause entomologists, agriculturists, and others the greatest concern.

With but a limited number of crops available, the spectrum of insect species in most agricultural ecosystems is relatively narrow. In most, there are four general classes of plant-eating insects: (1) Serious perennial pests; (2) intermittent but sometimes serious pests; (3) insects that are rarely if ever pests; (4) nonresidents of the ecosystem that enter it periodically for short times, and in so doing become serious, intermittent, or rare pests (Chant, 1964). While Group 1 is of major concern, the pests of Group 2 may be of equal importance to the plant breeder in that they may detract from the yield stability or crop potential of new varieties. Group 3 becomes a problem if a highly susceptible variety is introduced or cultural practices are modified.

AUTOREGULATION

Autoregulation or some form of self-determination of insect populations has been proposed by several entomologists. The role of behavioral patterns in this approach was discussed by Wynne-Edwards (1965). The social insects, which have developed an intensive colony pattern of living, do not show major outbreaks since cultivation is usually highly detrimental to the colonial way of insect life and such insects are reduced as soon as cultivation begins. In contrast, cannibalism in corn earworms and some other insect pests may curtail population densities in an individual corn ear or cotton boll, but does not prevent the earworm from being a major problem on corn and cotton in the United States.

Pimentel (1961) proposed a feedback mechanism of population regulation involving density pressure, selection pressure, and consequent genetic changes. This idea certainly gains credence when one considers the insect-host plant interactions and the physiological ramifications that occur over time in both insects and host plants.

DENSITY

When insects become crowded a whole series of factors becomes important. The natural enemies are greatly encouraged, and parasites, predators, and disease will often sweep the populations. In these instances the insects become hosts and follow patterns of epidemiology similar to those of green plants when *they* are the hosts.

Dense populations soon reduce the food supply and face starvation, as often occurs in the periodic epidemics of gypsy moths in parts of New England.

VIGOR

The relationship between vigor of the insect and population density has attracted some attention during the past decade. Most of the evidence comes from studies of forest insects, but the idea should be investigated in similar situations in cultivated crops. Campbell (1966) made a significant contribution by demonstrating in the forest tent caterpillar the genetic basis for a change in vigor in relation to population density. A fairly simple genetic system apparently determines metabolic rate, vigor, flight potential, and damage potential.

WEATHER

The weather, of course, affects insect populations just as it affects populations of fungi, bacteria, viruses, or nematodes. The body temperature of insects is determined by the temperature of their environment. Moisture regulation is a demand that must be met by all small animals since the volume-to-surface ratio is so small, but many species of insects have developed modifications in their metabolism to cope with extreme conditions. In the same way that climate and weather determine crop distribution and production, they also influence insects, but one cannot assume that what is optimum for the crop is optimum for the crop pest.

Endemic problems with grasshoppers are generally associated with areas having less than 20 in. of annual rainfall. Spider mites and some species of aphids appear to be encouraged by dry weather, but this may be a secondary effect, because the pathogens to which they are subject are adversely affected by low humidity, or because of a greater concentration of nutrients in the plant sap on which they feed. Adequate moisture during the growing season generally favors such chewing insects as cutworms and armyworms.

Cold winters also limit many insects, earworms and greenbugs, for example. Others, such as the European corn borer and wheat stem sawfly, can survive almost as far north as their hosts are grown.

HOW INSECTS FIND THEIR HOSTS

All parasites must travel from already exploited hosts to new ones, from affected plants to healthy ones; insects must move, too. Most disease organisms move passively by rain action or wind, whereas insects can fly or crawl. Thus plant pathologists tend to say that disease organisms "spread," while entomologists say that insects "migrate." Insects can therefore compensate for their relative scarcity in comparison with microbes because they have an added degree of freedom: they can fly, run or crawl in search of a host.

Only brief consideration can be given to host finding at this point. It may result from chance movements, but generally is made up of a series of events involving random movement coupled with oriented movements or other responses to the microclimate, followed by responses to mechanical and physical aspects of the host surface, acidity, odors, color, and taste factors of the host plant. In fact, for any significant infestation to occur, the host must elicit or permit continued feeding by the insect, a phenomenon characteristic of microbes also.

Insect migration and dispersal is more complex than is the movement of fungal spores (see Johnson, 1969). Some insects migrate in response to overcrowding or food shortages; others may migrate away from entirely adequate food sources and arrive at another location where food is not as readily available or in as good condition. The prime causes of migration have still to be identified for particular species. One must recognize that the propensity of an insect for dispersal determines the population density in a given crop unit and any factor in a plant variety that induces dispersal obviously reduces the insect population level in the immediate area.

Studies of field populations of plant-feeding insects show that the mortality rate is usually highest when the eggs are hatching and the young larvae are seeking to establish themselves on food plants and to obtain their first meal. The varied developmental patterns of insect species contribute to the complexity of host finding. In the simplest case, for example in aphids, the young and adults feed on the same host plant, and the eggs or nymphs are deposited where the female is feeding. Host finding is necessary only during periods of dispersal or when the original food host is depleted. Increasing stages of complexity are illustrated as follows.

• Grasshopper eggs are laid in the soil. The nymphs and adults feed on the same host plant so that once the nymph has found an appropriate host little additional effort is needed.

• In the corn earworm the female generally lays her eggs on a host appropriate for the young larvae. However, because the adults feed on nectar and the larvae have chewing mouth parts, the adult must seek another host.

• The most complex series may be illustrated by the common stalk borer, which lays its eggs on plant trash. When the larva hatches in the spring, it must crawl about in search of a small stemmed plant to begin feeding. After several growth stages it moves to a larger stemmed plant to complete its development and pupate. The adult feeds on nectar of flowering plants.

The second and third examples are illustrative of insects with distinct egg, larva, pupa, and adult stages, in which the larval stage is the major growing and feeding stage and the adult stage is basically a reproductive and dispersal phase in the life pattern.

Thornsteinson (1960) postulated two major types of chemical influences on host plant preferences. In the first instance chemical stimuli are present in virtually all plant species, but some plants are not consumed because they also contain an inhibitory substance. Inhibitory chemicals are found in a variety of plants. Some may be distributed at random among the plant groups; others occur in most members of certain taxonomic groups. In the second instance, feeding and oviposition are induced by specific chemical stimuli that are either limited to such natural plant taxonomic groups as Cruciferae, which contain stimuli for the cabbageworms; or the stimuli may be distributed at random. In several examples where chemical stimuli have been investigated thoroughly, frequently more than one substance was involved. Sometimes both attractants and repellents are present.

Environments with harsh, cold, or dry climates support relatively few species of plants and animals; environments with equable climates support enormous numbers of species. Where relatively few species are present each tends to be represented by large numbers of individuals. In the more equable environment, many species at each consumer level dissipate the energy, but each has a low population density. Hence, species abundance and diversity are related to the problem of balancing numbers of species against species density. In communities with relatively few species the fluctuations seem to be great, whereas where there are many species, fluctuations in population density are relatively small.

It follows that a system with a less complex plant species diversity has larger numbers of each species available. More effort should therefore be made to maintain diversity in these large populations while avoiding catastrophic fluctuations. This is equally true for plant species and for the insect pests. Therefore, as large areas are devoted to crop monoculture, the insect species that thrive on this monoculture have a broader genetic base for adapting to control strategies.

While the problems raised by host plant monoculture are well known, recent advances in weed control have still further reduced host plant diversity within the crop field. This simplification of the ecosystem is something that entomologists and agriculturists must bear in mind in deriving theoretical models for future food production.

COMPUTER SIMULATION

Just as a computer can be used to model the complexities of factor interaction in disease organisms, and thus to predict the course of an epidemic, so models of insect populations can be programmed for study. Watt (1964) has been a leader in developing simulations of insect outbreaks, which are mostly concerned with insects that attack trees. In the past, life tables (a compilation of all mortality factors affecting a population in a lifetime) have been considered necessary prerequisites to the development of models, but critical path analysis in conjunction with multiple regression may provide a shortcut in estimating variables.

On annual crops, where chemical pest-control decisions are generally made for individual fields, the greatest use of computers has been in deriving curves for estimating economic thresholds by sequential sampling. The Plant Protection Division of the United States Department of Agriculture makes annual predictions of the grasshopper potential in rangeland. The extent to which such work can expand depends on obtaining more support for the collection of basic data.

SIGNIFICANCE OF HOST UNIFORMITY

Other parts of this report emphasize the fundamental importance of host uniformity or, conversely, the lack of genetic diversity in the induction of epidemics. Both microbes and insects often respond to genetic uniformity of the host by producing biotypes or biological races. A biological race can be defined as a population within a species that is usually isolated to some extent by food preference, or that shows some distinct physiological characteristic but few corresponding morphological differences. It is distinguished from a subspecies or geographical race in that there need be no geographical isolation. The term biotype has been used most frequently with respect to biological races that are able to feed and survive on an otherwise resistant variety of a domesticated crop. The extent to which biotypes represent the transient stage in the evolution of insect species is most readily seen in crop plants, where plant breeders have made major changes within the crop or have so changed the plant species complex that insects have a better opportunity to express their preference and their ability to feed on different host plants growing in climates favorable to the insects.

The best known example of an insect biotype in relation to host resistance is the Hessian fly. Eight biotypes are now known to exist in the laboratory; six are found in the field. In the case of the pea aphid, entomologists have found, on a series of peas and alfalfa varieties, a series of biotypes with different survival and fecundity rates; but in this instance the effect on breeding work does not appear to be as deleterious as that of the Hessian fly. Several biotypes of the spotted alfalfa aphid are known in relation to the host plant, plus at least one biotype for resistance to the insecticide parathion.

There are three known biotypes of the greenbug, one of which has severely retarded breeding for greenbug-resistant wheats. The third biotype is able to thrive on grain sorghum and has become a major problem on that crop in the last few years.

In contrast to these instances, the Winter Majetin apple was reported to be resistant to woolly apple aphids in 1831, a resistance that has not broken down throughout the time that this variety has been grown commercially. Even though the variety is genetically uniform it has not been especially vulnerable.

The origin of biotypes in insects is not nearly as well understood as it is in plant disease fungi. The gene-for-gene concept developed in work on cereal rust diseases is most nearly approached in the case of Hessian fly-resistant varieties, but the biochemical basis has not been explained. Furthermore, all of the cases of biotype formation listed above involve insects with piercing or sucking mouth parts. Some earlier examples of race formation in nature were found in wood-boring insects. Here there appeared to be a conditioning of the female to lay eggs on the same host species as she had fed on in the larval stage. There have also been studies of ermine moth adults and larvae that show a preference for either apple or hawthorn. Codling moth populations may also show a preference for either pear or apple orchards. It is too early to speculate on specific versus general resistance concerning insects, since not enough genetic information is available.

INSECTS AS DISEASE VECTORS

Many insects do more damage as vectors of virus diseases of plants than they do by feeding on the plants. The efficiency with which an insect can transmit virus, whether or not the virus multiplies within the insect, and how long the virus remains viable on the mouth parts of the insect, varies from one insect-host-virus complex to another. Notable success on the control of curly top in sugar beets has been achieved by reducing the potential of leafhoppers as a vector by modifying the overwintering habitat. Where the insect acts as a passive agent for the virus, the population build-up of the insect is of major importance in determining the level of infestation on the host crop. While the probability of achieving virus disease control by breeding for resistance to the vectors would seem to be low, there have been successes on both rice and potatoes, and further work should be encouraged.

INSECT QUARANTINES AND SURVEILLANCE

The Plant Protection Division of the USDA is charged with the responsibility of recording insect infestations in the United States, as well as being on the lookout for new and introduced pests. This latter problem is particularly significant when one considers that such important insects as the Hessian fly, pink bollworm, European corn borer, cereal leaf beetle, Japanese beetle, spotted alfalfa aphid, and greenbug have all come to the United States from other countries and that more intensive surveillance might have prevented or delayed their entry. Commerce today is worldwide, and the potential for effective quarantines is dependent upon increased support for this activity in proportion to the steady increase in travel and shipment of commodities throughout the world.

Other countries are also working on programs for ascertaining insect infestation levels. In Japan, for example, 10,800 part-time staff members are involved as disease and insect control agents and assist in predicting infestations.

In the United States each state has a state insect survey coordinator, but the major effort is by 27 survey entomologists (in as many states) whose job is to gather information on infestations throughout the state. While his efforts have to be very diversified, he does have the help of a number of individuals in the state departments of agriculture, entomology departments, and other regulatory agencies. He usually has a fair estimate of any significant outbreaks. To the extent that chemical controls can be used in damping insect population increases, outbreaks do not go beyond the economic threshold as rapidly or frequently as was true before organic pesticides were available.

REFERENCES AND SUGGESTED READING

Campbell, Ian M. 1966. Genetic variation related to survival in Lepidopteran species. Breeding pest-resistant trees. NATO/NSF Symp. (Pennsylvania State Univ, August 30 to September 11, 1964). Proc. Pergamon Press, Elmsford, N.Y. 505 pp.

Chant, D. A. 1964. Strategy and tactics of insect control. Can. Entomol. 96: 182–201.

Geier, P. W. 1966. Management of insect pests. Annu. Rev. of Entomol. 11: 471–490.

Johnson, C. G. 1969. Migration and Dispersal of Inspects by Flight. London, Methuen and Co. 763 pp.

Morris, R. F. 1959. Single-factor analysis in population dynamics. Ecology 40:580–588.

Pimentel, David. 1961. Animal population regulation by the genetic feedback mechanism. The Am. Nat. 95:65–79.

Stern, Vernon M., Ray F. Smith, Robert van den Bosch, and Kenneth S. Hagen. 1959. The integrated control concept. Hilgardia 29:81–101.

Thornsteinson, A. J. 1960. Host selection in phytophagous insects. Annu. Rev. of Entomol. 5:193–218.

Watt, K. E. F. 1964. The use of mathematics and computers to determine optimal strategy and tactics for a given insect pest control problem. Can. Entomol. 96:202–220.

Wynne-Edwards, V. C. 1965. Self-regulating systems in populations of animals. Science 147:1543–1548.

CHAPTER 7

Economics of Epidemics

Contents

ECONOMIC AND SOCIAL INCENTIVES
   LEADING TO GENETIC SPECIALIZATION   83
CROPS GROWN   84
POTENTIAL ECONOMIC AND SOCIAL
   IMPACT OF EPIDEMICS   86
   Effects in Countries Where Food Is Scarce   88
   Effects in the United States   88
MEANS OF MODERATING ECONOMIC EFFECTS
   OF EPIDEMICS   91
   Reserve Stocks   91
   Reserve Capacity   92
   Crop Insurance   92
   Choice of Alternatives   93

81

In Chapter 3, Principles of Epidemics, we described the disease triangle—host, weather, parasite—and spoke of these as the driving forces of epidemics. However, economics also plays a part. The search for better crop varieties is motivated by man's desire to satisfy his economic wants. In a market economy, numerous pressures act on farmers, seedsmen, and processors to find and use varieties that are in some sense "best." In centrally planned economies, similar incentives exist but may take different forms. Frequently, specialization on only a few crop varieties is the result.

Thus, circumstances are encouraged in which the disease triangle may produce epidemics that destroy large portions of affected crops. The potential impact on society, however, depends on the kind of agriculture the nation has and on the role of individual crops in that agriculture. Means of dealing with epidemic risks include not only reduction of genetic vulnerability but also such devices as reserve stocks and crop insurance. Each alternative involves costs of one kind or another. Policy for dealing with epidemic risks may appropriately combine different alternatives and will vary with crops and economic circumstances.

ECONOMIC AND SOCIAL INCENTIVES LEADING TO GENETIC SPECIALIZATION

People desire more food, better food, and food at lower cost. These are the most pervasive reasons behind the crop improvement efforts that in many instances have led to increased vulnerability to epidemics. Virtually all countries that are seriously trying to improve the well-being of their people are making at least sketchy efforts to improve crop yields. Success along these lines is imperative in poor countries with rapidly growing populations. The United States has already achieved much; crop production per acre has doubled in the past 40 years. Had this not been so, food would now be less abundant and much higher priced than it is. Moreover, the scientific knowledge that produced the technological advance has great potential for relieving food scarcity in less-developed countries.

In the United States, both public policy and the operation of markets have stimulated crop improvement. Government support for a broad agricultural research program reaches back nearly a century. In more recent times, private firms have been doing an increasing proportion of all agricultural research and development. The market provides the final test of whether new crop varieties, however developed, are superior to old varieties; survival of the fittest in this case has economic as well as biological meaning.

Consumers' likes and dislikes regarding taste and appearance greatly influence the marketability of different varieties of crops. Processors and distributors may have additional preferences regarding storability, ease of transport, and processing characteristics. Uniform quality and appearance are often emphasized because of their importance in mass processing and merchandising. Vegetable processors ordinarily determine varieties to be planted on large acreages and insist on varieties tailored to suit their particular combination of requirements.

Farmers, of course, want varieties that have desirable market characteristics. In addition, they want high yields, adaptability to mechanical harvesting, ability to withstand wind and rain, and resistance to drought, insects, and diseases. Even aesthetics may be involved: for example, corn growers' preference for single-cross hybrids over the less expensive double crosses is probably due in part to the "eye appeal" of the highly uniform appearance obtained from single-cross seed. Finally, seed-producing companies want varieties that satisfy their customers—farmers and sometimes processors—and for which seed can be produced at the least cost consistent with maintaining desired characteristics.

It should be noted that "market superiority" of a crop variety in this context does not necessarily mean best quality from any single standpoint. Varieties judged to have best flavor or texture by persons familiar with the product may lose out to varieties more saleable because of attractive color, uniformity, or other visible characteristics. Varieties that would be preferred by consumers at equal prices may not be competitive because high growing, shipping, or processing costs make them expensive. Nutritional differences among varieties of the same crop may be given little weight because of consumer indifference, lack of consumer information, or difficulty in distinguishing among varieties in marketing the crop. Price premiums and discounts do exist, of course, for many varieties and grades of products, but they commonly represent market consensus or compromise rather than a single viewpoint. In many instances differentiation of quality in the market does not extend to every characteristic important to someone.

The numerous desired attributes in a crop variety greatly constrain plant breeders in their search for parent materials. At a given time only a very few varieties—and these may be closely related—are likely to combine desired characteristics with approximately equal effectiveness. Thus, the attributes that society wants in crops in order to have more, better, and less costly food and fiber strongly tend to narrow the genetic base of each crop and to increase its vulnerability to epidemics.

Market incentives strongly motivate crop growers and seed companies to avoid crop varieties that have been demonstrated susceptible to serious diseases. An example is provided by the vigorous efforts to increase supplies of seed resistant to Southern corn leaf blight after 1970. But the mere possibility of an epidemic because of a new or obscure disease is not likely to cause growers and seed companies to forego best-performing varieties. Moreover, the individual grower or seed company has no assurance that an alternative variety will be immune to whatever diseases appear in the future. Increasing vulnerability to epidemics is not likely to automatically generate self-correcting tendencies in the marketplace.

CROPS GROWN

Agricultural crops provide most of the world's food and much of its fiber. Some—wheat, rice, potatoes, cassava—are used directly for food.

Others—corn, alfalfa, and sorghum—are used in developed countries mainly or entirely as feed for livestock and poultry; the final food is, then, red meat, poultry meat, a dairy product, or eggs. The distinction between food and feed crops is only approximate, however, for all grains fed to livestock can be used for human food and in some parts of the world are staples in human diets, whereas food grains like wheat can be, and sometimes are, used as livestock feed.

Not all food comes from cultivated crops. Some livestock graze uncultivated hilly or arid land, and fish are an important part of the food supply in many countries. On the other hand, crops like cotton, jute, and tobacco are grown partly or entirely for purposes unrelated to food.

In many of the less-developed countries of the world, population presses heavily on the arable land, and a low level of technology is used in producing food in cultivated areas. As a result, crops must be used directly as food in order to stretch the supply of calories as far as possible. Little can be spared for livestock, which subsist mainly on forage that does not supplant food crops. Heavy reliance must be placed on the food crop producing the most calories per acre under local soil, climatic, and other conditions. Thus, wheat, rice, cassava, or some other crop is absolutely vital to those living in many parts of the world.

By contrast, the United States has both an abundance of agricultural resources and a high level of agricultural technology. Most of the land under cultivation grows crops used wholly or in part for livestock and poultry feed. The land produces far fewer calories for human consumption than it would if used for direct-consumption grain, potatoes, and similar crops. The meat, poultry, dairy products, and eggs produced from feed crops, however, much improve the quality of peoples' diets. Food grains, fruits, vegetables, and other direct-consumption crops, as well as cotton and tobacco, are produced in ample supply. The comparatively high incomes of most Americans and the great productivity of their agriculture permit them to enjoy a high level of living in terms of food.

Most of the other industrially advanced nations of the world either have productive agricultures of their own or earn sufficient foreign exchange to import food and feed needed to sustain a fairly high level of food consumption in terms of both quantity and quality. Western Europe, for example, is agriculturally productive and also imports feed grains and soybeans for its livestock and poultry.

Tables 1 and 2 show the relative importance of food and feed

TABLE 1 Use of Land (in Millions of Acres) for Leading Crops, by Areas of the World, 1968 (Approximate)

Crop	Europe^a	North and Central America	South America	Asia^a	Africa	Oceania	USSR	Mainland China^b	World
Wheat	71	86	21	102	21	27	166	68	562
Rice	1	4	13	219	9	c	1	80	327
Rye	20	2	1	2	c	c	30	c	55
Barley	38	19	3	28	14	3	48	32	185
Oats	19	25	2	1	1	4	22	6	80
Corn	28	82	39	39	42	c	8	24	262
Millet and sorghum	c	17	4	100	73	c	8	72	275
Total cereals^d	181	237	83	493	161	35	288	279^e	1,757
Potatoes	19	2	3	3	1	c	21	7	56
Sweet potatoes and yams	c	c	1	4	8	c	f	f	40
All pulses	14	9	11	67	19	c	12	23	155
Dry beans and peas	(10)	(8)	(10)	(24)	(5)	(^{c})	(8)	(13)	(78)
All oil seeds	7	62	19	77	28	c	25	60	278
Soybeans	(^{c})	(41)	(2)	(4)	(^{c})	(^{c})	(2)	(33)	(83)
Peanuts	(^{c})	(2)	(2)	(21)	(14)	(^{c})	(^{c})	(4)	(43)
Cotton	1	13	7	28	11	c	6	12	78
All crops^g	368	625	220	827	504	106	553	270	3,473

Source: FAO, 1970. Adapted from Production Yearbook 1969, V. 23.

^{a} Mainland China and USSR not included in Asia or Europe.
^{b} Rough estimates or old data. All crop total is for 1954, probably understated.
^{c} Less than 0.5 million acres.
^{d} Includes small areas of cereals not listed separately.
^{e} Contains statistical discrepancy of 3 million or more acres.
^{f} Not given separately but included in world total.
^{g} Arable land and land under permanent crops, some of it double cropped.

crops in the United States and in the world as a whole. Cereals collectively occupy about half of the world's cropland. Wheat and rice account for about half of the world's cereal acreage, and the other grains are used for food in many areas. In the United States, however, feed crops dominate by a wide margin. (It should be noted that acreage is not necessarily a good measure of economic importance; for example, the value of an acre of vegetables usually is far greater than the value of an acre of wheat or oats.)

POTENTIAL ECONOMIC AND SOCIAL IMPACT OF EPIDEMICS

Effects of a crop loss from disease, insects, or other cause obviously depend upon the severity of the loss, its duration, the importance of

TABLE 2 Acreages of crops harvested in the United States, 1970

Crop	Area Harvested (million acres)
Crops used principally for food^a	
Wheat	44.3
Rice	1.8
Rye	1.5
Sugarcane and beets	1.9
Commercial vegetables	3.2
Dry beans and peas	1.7
Potatoes and sweet potatoes	1.6
Peanuts	1.5
Fruits, tree nuts, farm gardens	7.0
Total food crops	64.5
Crops used principally for feed^b	
All corn	66.4
Oats	18.6
All sorghum	16.8
Barley	9.6
All hay	63.2
Soybeans	42.4
Total feed crops	217.0
Other crops^c	
Cotton	11.2
Tobacco	0.9
Flaxseed	2.9
Total other crops	15.0
All crops^d	296.5

Source: USDA.

^{a} Some wheat and by-products of food grains, peanuts, and sugar crops are fed to livestock.

^{b} Some feed grain is used for food or alcohol. About 35 percent of the value of the soybean crop is attributable to oil, and some soybean protein is used directly as food.

^{c} Cottonseed and linseed meals are livestock feeds; cottonseed oil is a food oil.

^{d} Includes small acreages of minor crops. Because of double cropping, the total cropland harvested was about 4 million acres less than the total for individual crops.

the crop, its role in the agricultural economy, and still other circumstances. Factors determining the extent and duration of epidemics are largely biological and environmental and are discussed in other chapters. The impact of a particular loss may be modified by stocks on hand or by the availability of reserve production capacity. This will be discussed in a later section.

EFFECTS IN COUNTRIES WHERE FOOD IS SCARCE

Severe damage to a crop that is the mainstay of the diet of people who normally consume only enough calories to sustain life obviously can have tragic consequences. The threat of famine will be greatest if food stocks are low at the outset, if the crop loss endures for more than one year, if a large area is affected so that nearby supplies cannot be obtained, and if food from abroad is not available. The Irish potato famine of 1845 is a classic example. The failure of the monsoon rains in India in 1965 and 1966 also caused great distress and might well have created famine had food aid not been provided from abroad, principally from the United States. Such problems often are compounded by poor transportation and poverty in the areas involved.

The less-developed, overpopulated countries are especially vulnerable—perhaps particularly so if losses occur when they are engaged in improving their agriculture. Often the crops grown before development began were low yielding but had varied genetic backgrounds and considerable resistance to local diseases and pests. Thus, they were not very vulnerable to epidemics. With development underway, however, one or two varieties introduced from another area may occupy most of the acreage. A new or formerly unimportant disease can then sweep virtually unchecked through most of the crop.

Economic and social effects of crop losses in less-developed countries are not confined to hunger alone: the country's meager supply of foreign exchange and limited credit may be exhausted in efforts to import food; political instability is likely to result from serious food shortages; and economic development in both the agricultural and industrial sectors may be greatly set back.

EFFECTS IN THE UNITED STATES

The impact of an epidemic in the United States is likely to be less tragic but more complex than in the case described above. Extreme damage over a period of several years to corn, the leading grain in the United States, would be perhaps the most devastating occurrence. The immediate impact would be softened by use of feed grain stocks carried over from earlier years (as happened in the corn leaf blight episode) and by returning to production any land previously withheld from cultivation. Other feed grains could be grown in place of corn but would yield lower tonnage in most locations. Large amounts of wheat could be fed to livestock.

The domestic shortage of corn could be partially alleviated by reducing exports of feed grains, wheat, and even soybeans, all of which are normally shipped abroad in large quantity. But the greatest potential adjustment to a shortage of feed grains would be to reduce the livestock and poultry populations. The first impact of liquidation of herds and flocks would be increased marketing of meat and poultry, and a reduction in consumer prices. After periods ranging from several months to more than a year, however, market supplies of particular meat, poultry, and dairy products would diminish and prices would rise sharply.

According to this model, price changes would be the principal means by which adjustments to a severe, long-lasting feed grain shortage would be brought about. The difficulty of curbing general inflation in the economy would be multiplied as cost-of-living increases were made in industrial wage rates. Families on fixed incomes, and especially low-income families, would be squeezed by rising food prices.

High grain prices in the United States would reduce exports of feed grains, wheat, and possibly soybeans. Total dollar earnings from exports would be likely to drop despite higher prices, and imports of some foods would rise. The balance of international payments would be adversely affected. Long-run damage to foreign markets for American farm products could result from temporary withdrawal.

Large income shifts within agriculture itself would be created in the event of devastation of the corn crop. Producers of meat animals and dairy and poultry products would be severely affected by high feed prices and forced to reduce production or to quit entirely. Total income received from production of all feed grains and wheat would probably rise, but corn growers who lost their crops would be hard hit, while other grain producers would receive windfall gains. Once the epidemic had passed, the shifts of income as normal patterns of production were restored would be in the opposite directions.

Groups other than farmers would also be directly affected by a serious corn shortage and consequent reduced production of livestock and poultry products. Businesses engaged in handling or processing these products would lose volume and profits, and some of the labor they normally employ would lose their jobs. (Corn is not a product for which much hired farm labor is employed, but loss of a hand-harvested fruit or vegetable crop would have an immediate impact on employment of farm workers.)

The consequences of moderate damage to corn, or of severe damage to a less important feed grain, would tend to be similar but less drastic. Whether the loss was severe enough to cause significant reduction in livestock output would be critical in determining the extent and kind of impact.

Similarly extensive and prolonged damage to the soybean crop would have serious consequences for the economy. The principal protein feed for livestock and the leading source of vegetable fats and oils would be in short supply; one of our principal farm exports would be hard hit. Substitution of other protein feeds and of other fats and oils for those derived from soybeans would ease the situation somewhat, but soybean products so dominate this complex of commodities that a substantial loss of the crop would have far-reaching effects.

Wheat provides one-sixth of the food energy in the average American diet. The most important class, hard winter wheat, accounts for less than half of this, and large exports create a considerable cushion of safety for American consumers. However, curtailment of exports because of crop loss would tend to reduce dollar earnings. It might also significantly impair the food supply available to a less-developed country that had been partially dependent on American wheat made available to it on less than commercial terms under Public Law 480 (Food for Peace). Except for this possibility, loss to any class of wheat due to disease would have less far-ranging effects than similar proportionate losses of corn or soybeans.

The potential economic impact of losses of other crops, occasioned by disease or insect infestations, can be inferred from what has already been said. Losses usually will be serious to growers whose crops are damaged and may be important to nonfarm businesses and to labor handling the crops. Consumers will pay higher prices. The national impact will be slight if the product is a minor one or is replaceable in consumers' diets. It will be more significant if the product is a major or irreplaceable one.

The American economy, including agriculture, is highly interdependent and requires continuity, and regularity for effective functioning. Though danger of famine as a result of epidemics in crops is remote, severe and prolonged damage to a major crop can be highly disruptive to agriculture itself and can set in motion shock effects that extend to the retail price of food, the level of industrial wages, inflation of nonfood prices, the plight of the poor, and the international balance of payments. Prevention of crop losses and means of mitigating their effects are important to society as a whole.

MEANS OF MODERATING ECONOMIC EFFECTS OF EPIDEMICS

The probability of substantial loss of a crop because of an epidemic can be reduced in many instances—though never to zero—by modifying the crop's genetic make-up. However, other means can deal with the hazard of crop failure. These can supplement genetic modifications or substitute for them.

RESERVE STOCKS

Stocks carried forward from a previous year may be available as a buffer between approximately stable consumption and fluctuating production. In the less-developed countries of the world, producers of cereals or other storable crops ordinarily carry some of the production of an especially good year into the next, and small surpluses may also be carried forward by dealers. Some governments maintain stocks for emergency use. Thus modest reserves may be—but are not always—available if an epidemic strikes a nation's staple crop. The reserves will moderate the impact of the epidemic in some degree during the first year, but are unlikely to be available in subsequent years.

In the United States, risk calculations by farmers, by processing and storage companies, and by speculators lead to private storage policies that more or less reflect calculable risks of crop losses from weather and other ordinary hazards. More important for a number of commodities, however, have been large stocks held directly or indirectly by the federal government as a result of its farm price-support policies. Such stocks often have much exceeded the amounts that the private trade would have held. (In light of government stocks, privately held stocks have often been especially low.) Government stocks have provided more protection against shortages than would otherwise have existed. Their availability considerably reduced the adverse impacts of blight damage to the corn crop in 1970.

The probability of an epidemic in a given storable crop within a given span of a few years is ordinarily too low to enter the private calculations that determine the average level of privately held stocks. If the government determined, however, that additional stocks were needed for reasons not entering private calculations, it might logically carry larger stocks as a matter of national policy. Epidemics are not the only hazard that require decisions of this type. Rare droughts of the magnitude of the Dust Bowl days, the possibility of unexpected needs for food abroad, and national defense are other examples. Thus epidemics are one of a class of uncertain events that might be dealt with by storage policy.

RESERVE CAPACITY

Success in increasing per-acre yields of crops in the United States has reduced the acreage needed for crop production. The land thus made available could be used to offset moderate production losses due to epidemics or to grow substitute crops if they are available. Thus the productive power generated by crop technology has enhanced agriculture's ability to deal with risks associated with that technology.

Ordinarily, land released by crop improvement would not be readily at hand to resume production. Under the farm programs of the past 10 or 15 years, however, a large portion of released land remains in operating farms and can quickly be returned to cropping. Whatever the merits of the programs in other respects, such land constitutes reserve capacity that can be used to compensate for reduced crop yields or to meet unusual needs.

CROP INSURANCE

The individual grower who loses his main crop because of an epidemic suffers a severe financial setback whether the crop is a major one in the nation's economy, or a minor one. Insurance is one form of protection: the federal government operates a limited crop insurance program under which losses due to causes beyond the grower's control, including disease, can be covered. Some private companies have small or experimental programs covering losses from disease, but growers often consider crop insurance (except hail insurance in some locations) expensive and are slow to use it if the danger of an epidemic appears remote. On the whole, it seems unlikely that insurance against unforeseen epidemics will be widely used by growers unless serious disease outbreaks occur more frequently than in the past.

CHOICE OF ALTERNATIVES

Every means of dealing with losses from potential epidemics in crops entails a cost. Introducing greater genetic variability in cultivated crop varieties is likely to reduce per-acre yields to some extent and perhaps their acceptability in other respects. Someone has to pay the costs if extra stocks, or reserve cropland, are maintained. Insurance premiums are costs. No alternative gives complete protection: genetic vulnerability cannot be reduced to zero, reserves may not be large enough to meet all contingencies, and crop insurance is unlikely to cover all the expenses, even of the insured.

A program of genetic modification, if practicable, would give protection to consumers, growers, and all other groups, in both the short and long run. Reserve stocks protect consumers, livestock producers, and processors in the short run, but not crop growers. Reserve land provides long-run protection to all groups against moderate reductions in per-acre yields. Insurance, even if feasible, would protect only growers.

The United States has several alternatives in dealing with the susceptibility of crops to epidemics. This is not true for most of the less-developed countries, which cannot readily carry large reserve stocks, have no excess cropland, and do not harvest large feed crop acreages that might be used to grow human food. Such countries necessarily rely heavily on genetic resistance in their food crops.

PART II

VULNERABILITY OF INDIVIDUAL CROPS

CHAPTER 8

Corn

Contents

GENETIC DIVERSITY	98
Trends in the Last 50 Years	100
The Rise of Inbreds	101
Surveys of Inbred Use	103
Improved Inbreds	105
Planned Diversity	107
The Cytoplasm	108
Benefits	109
WHY YIELDS INCREASED	110
PESTS AND DISEASES	112
RISKS ENHANCED BY GENETIC UNIFORMITY	112
GERM PLASM RESOURCES	113
BROADENING THE GENETIC BASE	114
SUMMARY	117

97

GENETIC DIVERSITY

Corn (*Zea mays* L.) exhibits great genetic diversity in plant, ear, seed characteristics and diversity in resistance to diseases and insect pests, although the last is less well catalogued. Not all of the genetic diversity within the species is readily available within any given region because of differential response to day length, temperature, and other environmental factors. Several hundred races of corn have been described for the Latin American countries; few if any of these are directly useful in the continental United States and fewer still in the Corn Belt.

Most of the maize germ plasm of the United States Corn Belt is derived from mixtures of essentially two races, the northern Flints and southern Dents. This material, although possessing much genetic variability, nevertheless represents but a small fraction of the total genetic diversity within the species.

Genetic diversity may be considered in two separate categories—parental and regional. Some degree of parental diversity is basic to the expression of hybrid vigor at the commercial level. The genes re-

98

sponsible have not been individually identified, but it is generally assumed that large numbers are involved. Yield trials remain the sole method of identifying those rare genotypes that perform well in hybrids. First concern has been to maintain and enhance the type of diversity required for acceptable levels of hybrid vigor.

We shall be more concerned here with regional diversity, that is, diversity within and among populations of open-pollinated varieties, synthetics or composites, and double- or single-cross hybrids. In open-pollinated varieties pollen source is not controlled, so that no two plants of a variety are identical in genotype, and therefore upon inbreeding exhibit a high degree of heterogeneity among lines. In addition to diversity within the variety, additional diversity has been realized because of the large number of varietal strains or selections grown by farmers within any geographical subdivision.

Regional diversity changed rather drastically with the substitution of hybrids for open-pollinated varieties. At the single-cross level, all plants of a specific hybrid have a high degree of genetic similarity. In such a field, therefore, genetic diversity is at a minimum. Diversity within a geographical area can be achieved only through the use of many genetically distinct single crosses widely distributed over their geographical range. Such a distribution would provide diversity at the regional level but not in individual farmers' fields.

The first use of hybrids involved double rather than single crosses. Double-cross hybrids are more variable since they are produced by the union of the gene arrays from two different single crosses. Thus, double crosses provide some level of diversity within a single field and an area diversity that is probably substantially greater than any likely to be achieved through the use of single crosses.

Genetic diversity both within and among populations has long been recognized as a stabilizing influence. Restriction in genetic diversity arising from increasing dependence upon a single genotype or group of closely related genotypes carries an element of risk. A narrow genetic base provides a favorable genetic environment for the destructive increase of new races of pathogens or strains of insects. Uniform material is also less buffered against any unusual environmental challenge. This is true in general, but variety-year and variety-location interactions (the way such effects are measured) seem to be no greater for some single crosses than for the best double crosses. The effects appear as unusually high or low yield levels, extreme susceptibility to stalk lodging (as occurred in 1965) the leaf blight epidemic of 1970, or other aberrant behavior.

TRENDS IN THE LAST 50 YEARS

An examination of corn grown in the period from 1920 to 1970 discloses a decrease in genetic diversity that has accelerated since the adoption of hybrid corn. Detailed data are not available for the Corn Belt, or any other area of the United States, on the number of distinct varieties grown or the acreage devoted to each. Similarly, data are incomplete on the acreage devoted to individual hybrids on any given date or the shifts in hybrids over time. Lacking such detail, one must depend on historical patterns and survey data: evidence of changes in genetic diversity must be descriptive rather than definitive.

Extensive studies on varietal hybridization were conducted during the period 1877 to 1920. Richey (1922) has summarized data from 244 publications of this period. In general the largest yields were obtained from crosses between high-yielding varieties. Where parents were of the same kernel type and of similar climatic adaptation, intervarietal hybrid vigor did not often exceed 10 percent. Conversely, crosses of Dents with Flints, Dents with Flours, or Flints with Flours showed a consistently high level of hybrid vigor, in some cases equal to that observed in the first commercial double-cross hybrids. A high degree of genetic similarity was found within a given geographical or environmental zone, thus emphasizing the need for genetic diversity at the hybrid level and the importance of using high-yielding varieties as parents.

The *Iowa Corn Yield Test* was begun in 1920 to identify high-yielding strains or varieties. Entries were made by individual farmers who were seed corn specialists or enthusiasts, each on the conviction that his strain of corn was superior. The state test provided the first opportunity to compare differences in yielding ability among a large number of entries under uniform testing procedures. In view of maturity and adaptation differences, the state was divided into four sections each with three districts, and entries could be made on either a district or sectional basis. This scheme permitted each entrant's strain of corn to be tested in the area of its presumed adaptation. Sizeable differences in yield were observed (Table 1).

The demonstrated superiority of certain strains undoubtedly led to increased seed sales. This concentration may in turn have led to a small increase in yield potential but would have had only a small effect on genetic diversity within the state.

The *Iowa Corn Yield Test* later served as a vehicle for comparing relative yield capacity of varieties and hybrids. Similar but somewhat less extensive tests were conducted in many other states and undoubtedly influenced the selection of strains for inbreeding studies.

In the period 1920–1930, preceding the introduction of commercial hybrid corn, a relatively few open-pollinated varieties were used as foundation material for the development of inbred lines. In the central and southern portions of the Corn Belt, the yellow varieties used most extensively were Reid—or its many substrains or derivatives—Leaming, Krug, and Lancaster Surecrop. The most widely used white varieties were Boone County White and Johnson County White, which are related, and Pride of Saline. An equally restricted base was utilized in the northern portion of the Corn Belt and in the southeastern and southern states.

THE RISE OF INBREDS

Several states began inbreeding programs following Shull's (1909) report on inbreeding and hybridization as a method of corn improvement. Most of these were of short duration, and the majority of the public programs were started in the early 1920's. In each state the established local varieties were used as parental material. Only minimal selection was practiced during inbreeding because the prevailing idea of the period was that any inbred that could be maintained was potentially valuable. With accumulating experience it was realized that inbred lines to be commercially useful must satisfy two cri-

TABLE 1   Range in Yield for the Open-Pollinated Entries in the Iowa Corn Yield Test Conducted in 1926

District No.	Number of Items Tested	Yield in Bushels per Acre	
		Highest	Lowest
1	29	57.71	44.37
2	28	47.09	30.71
3	19	48.08	36.73
4	74	74.40	50.01
5	62	65.20	40.82
6	29	42.32	31.84
7	39	72.57	53.09
8	54	74.60	53.70
9	49	101.73	58.50
10	35	52.98	43.38
11	33	65.43	47.80

Source: Robinson, J. L., and A. A. Bryan. Iowa Corn Yield Test. Results of 1936 tests.

teria: (1) They had to be moderately vigorous and easy to maintain through production of acceptable yields of good-quality seed; and (2) they had to produce superior yields in hybrid combinations. With the realization of the importance of selecting for general vigor rather than performance in a cross, many lines were discarded during the inbreeding process. Lines surviving to the fourth or later generation of selfing were evaluated in single-cross combinations. Many lines were discarded because of unsatisfactory yield levels or liability to root breakage, stalk breakage, or disease susceptibility. At a later stage, a top-cross (inbred crossed with a variety) test was substituted for the preliminary single-cross evaluation, with considerable gain in efficiency. Single-cross or top-cross data were then used to predict double-cross performance. Only the hybrid combinations with a predicted superior performance were produced and subjected to replicated yield tests. The experimental combinations retained were substantially superior to the open-pollinated varieties for which standards had been established. The first double-cross hybrids that came into commercial use gave average yield increases of 25 to 30 percent. With the general acceptance of hybrids the best of the then current hybrids were chosen as the standard of reference for new experimental combinations.

The substitution of hybrids for open-pollinated varieties progressed most rapidly in the Corn Belt (Figure 1). Among the first hybrids produced commercially, none that came into extensive use had all component inbred lines derived from a single varietal source. Many

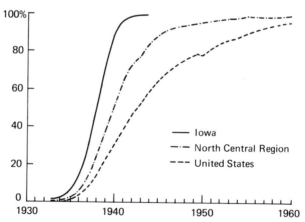

FIGURE 1   Hybrid corn adoption. Percent of corn acreage planted with hybrid seed.

of the high-yielding hybrids involved lines developed in two or more programs. This may have been due in part to genetic differences in the parental populations, but their use was certainly fostered by the widespread exchange of the better lines.

During and after the transition from open-pollinated varieties to hybrids the search continued for still better inbred lines ("better" with respect to performance at both the inbred and hybrid level). Some effort was devoted to a resampling of the original varietal sources. Because this approach was considered inefficient, greater attention was directed to: (1) Improvement of existing lines through the back-cross technique; (2) development of new lines from hybrid combinations; or (3) random mating of populations developed through recombining existing inbred lines (synthetics). The first two of these procedures involves a further restriction in the genetic base on which United States corn production depends. Furthermore, as elite lines emerge from breeding programs, they are frequently used as sources of improvement of other lines possessing one or more unsatisfactory characteristics.

SURVEYS OF INBRED USE

Much information on the progressive reduction in genetic diversity has come from three surveys that indicate the extent of commercial use of inbred lines developed, tested, and released by public agencies. In making the survey, a list of the publicly developed lines was supplied to the American Seed Trade Association (ASTA), and its members responded in terms of bushels or pounds of each type of seed produced. The first survey was conducted in 1956 and included all lines released during the period 1946 through 1955. Returns were received on 4.7 million bushels, approximately 40 percent of the 1956 seed requirement. Similar surveys were conducted in 1964 and 1970. In each case lines from the previous survey for which no commercial use was indicated were dropped from the listing. New lines released up to two years before the survey date were added. Response from the 1964 survey covered 6.3 million bushels and from the 1970 survey 6.0 million bushels. Thus, in each case, publicly developed lines were involved in the production of 40 to 50 percent of the total seed requirements.

Returns from the three surveys provide information on extent of use, but information of this kind cannot be related to the actual pedigrees of the commercial hybrids sold. There is reason to believe, however, that the inbred lines developed by companies representing the seed industry have essentially the same genetic background as the publicly developed lines, both having been developed from the same varieties. Any trends established for the public lines as a group, therefore, should also apply to the lines used in commercial hybrids.

Intense selection is practiced during both inbreeding and preliminary hybrid screening. Elimination is very rigorous during these stages. Selection continues during the extensive hybrid evaluation trials, and only a limited number of lines are retained. Of this group, only a very few are good enough to merit release. Inbred lines released by public institutions are available for use by any producer of hybrid seed, who may subject them to evaluation criteria different from those used by the originator. Many released inbreds are never used in commercial hybrids. Others are used to a limited extent; still others become widely used. This situation for each of the three surveys mentioned above is illustrated in Table 2. Study of the surveys justifies the following generalizations with respect to evaluation and acceptability:

• Evaluation—new lines must meet certain requirements for performance, imposed by the originating station, before becoming candidates for release. In spite of these requirements less than half of the released lines appear in the commercial use surveys (Table 2). Public lines are normally evaluated in combination with a limited series of testers under the environments prevailing within the state. The released lines must next meet more stringent requirements if they are to become commercially acceptable and useful. They will be evaluated in combination with different testers or seed parents under more diverse environmental or other challenges. They may be unacceptable on the basis of seed production capabilities, yield performance, har-

TABLE 2   Results from the 1956, 1964, and 1970 Surveys on the Use of Publicly Developed and Released Lines of Corn

| Year of Survey | Total Number of Lines | No. of lines in Following Classes | | | Sample response as % of total Seed Need |
		Use Exceeding 3.0% of Total Need	Use between 2.9 to 0.1% of Total Need	No Use Indicated	
1956	156	9	82	65	36.6
1964	249	13	103	133	55.8
1970	197	7	70	120	49.8

vesting characteristics, or susceptibility to some disease or insect pest; or they may exhibit no significant superiority over lines in current use.

• Acceptability—the lines that achieve some degree of acceptability may be utilized for relatively short periods (P8, R61, and Oh41) or may retain their usefulness over relatively long periods (Oh43, B14, and C103) (Table 3). It should be pointed out that several lines predating the 1956 survey were very widely used and some of these have continued in use to the present time (WF9, Hy, 38–11, M14, and others), both in their original state and as recovered or second-cycle lines. The data in Table 4 suggest a changing pattern in the extent of use of publicly developed lines. The samples listed include the most extensively used lines within each of the three surveys. In the 1956 survey six lines were involved in hybrids totaling 1.6 to 7.4 percent of the seed requirements. In contrast, each of the five most widely used lines indicated by the 1970 survey were used in hybrids representing 7.4 or more percent of the seed requirement. The fifth highest line of 1970 was used as extensively as the top line of 1956. The 25 most widely used public lines are listed in Table 4 together with their indicated usage.

IMPROVED INBREDS

As lines achieved widespread use their specific limitations became apparent, and extensive programs were set up to obtain improved versions through either backcrossing or, less commonly, second-cycle selfing

TABLE 3   Percentage Use of the 5 Most Extensively Used Public Lines in Each of the Three ASTA Corn Surveys

| Lines | Commercial Use as a Percentage of Total Seed Requirements for the Years Indicated | | |
	1956	1964	1970
C103	7.4	11.9	4.2
Oh43	5.2	15.7	11.7
B14	2.5	8.2	8.6
R61	1.6	1.0	0.0
P8	1.6	0.2	0.0
B37	0.0	2.0	25.7
Oh41	1.6	1.8	0.0
W64A	0.0	0.9	13.0
A632	0.0	0.0	7.4

TABLE 4   The 25 Most Widely Used Publicly Developed Lines of Corn Based on the 1970 ASTA Survey

Inbred Designation	Year of Release	Origin[a]	1970 Indicated Usage (bu)
B37	1958	Stiff Stalk Synthetic	3,091,132
W64A	1954	Wf9 × 187-2	1,559,140
Oh43	1949	W8 × Oh40B	1,407,087
B14	1953	Stiff Stalk Synthetic	1,032,326
A632	1964	Mt42 × B14[3]	891,761
C103	1949	Lancaster	509,552
A619	1961	A171 × Oh43[2]	413,362
A239	1957	A374 × A73	273,299
H84	1965	B37 × Ge440	243,759
CI.66	1958	L97 × K55[2]	225,803
Mo17	1964	187-2 × C103	209,428
CI.64	1955	L97 × K64[2]	159,839
C123	1961	C103 × C102	159,272
W153r	1953	Ial53 rec.	158,434
B14A	1962	Cuzco × B14[9]	150,564
N28	1964	Stiff Stalk Synthetic	140,898
A554	1960	Wf9 × WD[2]	123,362
A251	1957	A73 × Oh51A	121,597
W117	1963	643 × Minn.13	114,470
K55	1946	Pride of Saline	99,286
K64	1946	Pride of Saline	99,286
A635	1964	ND203 × B14[3]	81,440
B57	1963	Midland	70,672
MS1334	1962	Amargo × Golden Glow[2]	63,311
A634	1964	Mt42 × B14[3]	50,191

[a] Superscripts refer to the number of generations of backcrossing.

(the development of new inbreds from single- or double-cross hybrids). Of the 25 lines listed in Table 4, five represent direct isolates from open-pollinated varieties (C103, K55, K64, B57 and MS1334); three were derived from synthetics (B14, B37 and N28); seven involved single crosses as parental source material (W64A, Oh43, A239, H84, Mo17, C123, W117, and A251); and the remaining lines represent backcross derivatives. Such second-cycle lines may replace the original line, or both the original and backcross derivatives may remain in extensive use, e.g., K64 and CI.64, K55, and CI.66. Several backcross derivatives of the same original lines may become widely used. An extreme example of such a situation is illustrated in Table 5.

TABLE 5 The Extent of Use of B14 and B14 Backcross Derivatives as Indicated in the 1970 ASTA Corn Survey

Year of Release	Designation	Pedigree[a]	Commercial Use as Percentage of Total Seed Requirement
1953	B14	Stiff Stalk Synthetic	8.6
1962	B14A	Cuzco $\times$ B14^9	1.3
1964	A634	(Mt42 $\times$ B14^4	0.4
1964	A632	(Mt42 $\times$ B14^4	7.4
1964	A635	ND203 $\times$ B14^3	0.7
			Total 18.4

[a] Superscripts refer to the number of generations of backcrossing.

Line B14 ranked fourth in the 1970 survey. It was released in 1953 and had achieved third rank in the 1956 survey. The first of the B14 backcross derivatives was released in 1962 and others in the period through 1968. With the exception of B14A, each of these lines is somewhat earlier than B14, thus permitting their use in areas requiring hybrids that mature in a shorter period. The combined use of B14 and its derivatives totals 18.4 percent. Similarly the combined use of B37 and its relative H84, and Oh43 and its relative A619, total 27.7 and 15.2 percent respectively, of the 1970 seed requirements. The same general situation, but in a less extreme form, would hold for widely used lines of an earlier vintage: WF9, Hy, 38-11, and C103. Any new line of outstanding merit will be likely to follow the same general pattern.

PLANNED DIVERSITY

In 1947 the North Central Corn Improvement Conference allocated all of the lines released up to that time to one of two groups, A or B. All attempts to improve these lines, through either the backcross or second-cycle selfing approach, was to be restricted to a given group. The objective was to maintain parental genetic diversity, and therefore potential hybrid vigor between groups. This procedure has been followed with some success. The degree of utilization of certain lines, however, was not foreseen; and the group assignment procedure, while protecting hybrid vigor capabilities, does not protect against a growing genetic similarity among the hybrids currently grown.

Evidence of another kind illustrates the gradual reduction in genetic diversity among the widely used inbred lines. Of the 25 lines listed in

Table 4, only eight were derived from open-pollinated varieties or synthetics. C103 was isolated from Lancaster Surecrop; B14, B37, and N28 were derived from Stiff Stalk Synthetic; K55 and K64 from Pride of Saline, B57 from Midland, and MS1334 from the varietal hybrid (Golden Glow X Amargo) Golden Glow. Only three lines that have achieved significant commercial use have been derived directly from varietal sources within the past 10 years (B57, N28, and MS1334).

THE CYTOPLASM

Thus far we have been concerned with details indicating the progressive loss in the nuclear component of genetic diversity. The situation involving the cytoplasmic component is less clear. Before 1950 all hybrids involved cytoplasm that was derived directly from the open-pollinated parental source. Unfortunately current genetic procedures for characterizing cytoplasms are extremely primitive and are limited to such obvious characteristics as cytoplasmic sterility and various chlorophyll and growth patterns. It appears very unlikely that these striking changes represent all of the differences that actually exist among cytoplasms of various inbreds with normal (N) cytoplasm.

Mutations in nuclear genes arise from substitution, deletion, or addition of the four nucleotides of DNA. Inasmuch as both plastids and mitochondria have their own DNA, which is as likely to undergo mutational changes as is the nuclear DNA, it is reasonable to assume that the DNA of these two cytoplasmic organelles has likewise been modified by comparable mutations.

Where cytoplasmic inheritance has been studied in detail, nuclear-cytoplasmic interactions are involved. Certain functions of the cytoplasmic organelles are under control of the nucleus; others appear to be mediated by the organelle DNA-RNA system. If the assumption of differences among common cytoplasms is valid, then cytoplasmic diversity is decreased by the widespread use of a sterile cytoplasm that places all combinations of nuclear genes in a single common cytoplasm.

In the early 1950's limited commercial use was made of both Texas (T) and USDA (S) types of cytoplasmic sterility. Both types were used to an increasing extent until the late 1950's when the S type of sterility was gradually abandoned on account of its unsatisfactory level of environmental stability. The use of T cytoplasm continued to increase until 1968 when it was utilized in 75–90 percent of the commercial hybrids marketed. The T type of cytoplasm conditioned increased susceptibility to *Phyllosticta maydis* and Race T of *H. maydis*. It should be emphasized, however, that the substitution of T for N cytoplasm in commercial hybrids did not represent any drastic change in regional genetic diversity. Unfortunately the substituted type (T) exhibits susceptibilities to various developmental hazards not characteristic of the original normal type or types.

Extensive tests during the period 1960–1968 (Duvick, 1959; Grogan *et al.,* 1965) provided comparative data on hybrids involving N and T cytoplasms. The T cytoplasms were sometimes superior, particularly under conditions of stress. The fertility restoration gene, *Rf,* often led to small but consistent increases in stalk breaking (Russell and Marquez-Sanchez, 1966). As early as 1968 some of the seed companies observed that hybrids with T cytoplasm were apparently deteriorating in performance, both in reduced yields and an increased incidence of stalk breaking. This was particularly apparent in the northern portion of the Corn Belt where *Phyllosticta* leaf blight had become a problem. Similar observations were made in southern areas but were not related to any specific disease. As a result of these observations there was some shift back to N cytoplasm types before Race T of *H. maydis* was identified. Had it not been for these observations of deteriorating performance and the beginning of a shift back to N cytoplasm, losses in 1970 could have been of even greater magnitude.

BENEFITS

The progressive reduction in regional genetic diversity has brought immediate, sizeable, and continuing benefits, as well as nebulous and varied risks. New risks arise, in general, from either mutations in fungal or insect pests or from introductions of new parasites. Thus, neither the time nor the nature of the new risk is predictable. The acceptance of risk is normal in agricultural production.

The advancements in agricultural productivity that have occurred in the past 30 years are illustrated in Figure 2. Five-year moving averages have been used in this graph to minimize environmental effects and thereby emphasize the uniformity of the trend. By 1945 the transition from open-pollinated varieties to hybrids had been largely completed. Some have assumed that productivity changes since 1945 have their basis in production practices with little or no genetic input. A detailed examination, however, suggests that at least some of the

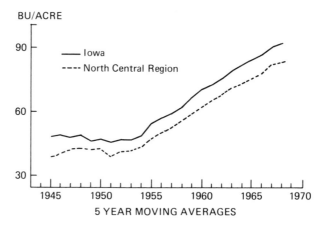

FIGURE 2   Corn yields, Iowa and North Central region, 1943–1970.

production practices depend on genetic input and either would not have been adopted or their rate of adoption would have been far different without it.

WHY YIELDS INCREASED

The main changes in agricultural practices that contributed to the post-1945 yield increased include: (1) A progressive increase in planting density; (2) an increased use of fertilizer, particularly nitrogen; (3) development of machinery for planting and harvesting designed to accommodate the new production systems; (4) increased use of herbicides and insecticides; (5) a transition to new hybrids and hybrid types conducive to economic benefits from other changes in production practices.

The number of plants per acre has increased at least 50 percent in the post-1945 period through changes in planting patterns: corn is now drilled or power-checked rather than planted in hills with the equidistant spacing characteristic of the check-row system. Row spacings have steadily decreased, the current norm being around 28–32 in. instead of the previous 40–48 in.

Prior to 1953, the average rate of nitrogen fertilization in Iowa was less than 10 lbs per acre. This rate had increased to 80 lb by 1965; by 1970, many of the more progressive farmers were using 150–200 lb of nitrogen per acre. Much of this has been applied as liquid ammonia, requiring either purchase of new equipment or contracting for custom application.

The use of both herbicides and insecticides has increased, as higher potential yields made their use economic.

Gradual change in production practices accompanied changes in hybrids and hybrid types. Unfortunately detailed data on such changes are available only for open-pedigree hybrids, not for proprietary hybrids. The data are expressed as acres of seed grown or pounds of seed produced under certification. Table 6 shows, for Iowa, the hybrid produced in greatest amount in four selected years and the amounts of the other three hybrids produced in the other three years. The rapidity and magnitude of the substitutions is very striking.

Certification played a key role in the early years of hybrid seed production. Since about 1945, however, a steadily decreasing percentage of the hybrid corn seed marketed has been certified. This tends to limit the applicability of the data in Table 6. Study of records of proprietary hybrids, produced under certification, indicates the same general substantive changes, so the trend itself is not subject to question.

More critical information on the magnitude and rapidity of genetic shifts is provided in Table 4 in the column "Year of Release." In this most extensively used group of lines, half were released after 1960 and only four lines (K55, K64, C103, and Oh43) were released prior to 1950.

The substitution of one hybrid for another had only a slight effect on regional genetic diversity. Some lines were intolerant of close spacing or failed to show desired response to increased rates of fertilization. They were replaced by others more suited to the changing technology. A change in hybrid type, involving the gradual replacement of double crosses with three-way or single crosses, had a much more drastic effect on genetic diversity, however. In 1955 little com-

TABLE 6   Seed Production of Selected Certified Corn Hybrids Expressed as a Percentage of Total Open-Pedigree Seed Production in Iowa for Years Indicated

Hybrid Number	Pedigree	1950	1955	1960	1965
Ia4297	(WF9 × I205)(187-2 × M14)	17.0^a	5.1	1.4	0.0
Ia4376	(WF9 × B6)(187-2 × M14)	0.0	13.8^a	10.2	0.0
Ia4570	(WF9 × B14)(187-2 × M14)	0.0	0.0	13.8^a	0.0
Ia4630	(WF9 × Oh51A)(B21 × M14)	0.0	0.0	8.8	17.1^a

^{a} Open-pedigreed certified hybrid produced in greatest quantity in the year indicated.

mercial use was made of single crosses, but the transition was fairly rapid, and by 1970 approximately 75 percent of the corn acreage in the Corn Belt was planted to single crosses. If all lines were used equally, the mere shift from double crosses to single crosses would reduce genetic diversity by one-half. With a disproportionate use of certain lines (Tables 4 and 5), the reduction in diversity becomes even more striking. Without these combined changes in technology, yields would doubtless have remained at or near the 1945–1954 levels.

PESTS AND DISEASES

Inbred lines developed or released during this transition period tend to exhibit moderate to high levels of resistance to the diseases common to the locality in which they were developed. The lines represent an unselected sample, however, in their reaction to either new parasites or insect pests or to new races of established parasites. Reaction to parasite or insect pests is also influenced by changes in technology, e.g., increased plant populations. Thus the identification of adequate host-plant resistance must be a continuing process. For further details on this topic, see Chapters 5 and 6.

RISKS ENHANCED BY GENETIC UNIFORMITY

Whenever large areas are planted to a single genotype, conditions become favorable for the rapid spread of virulent parasites and pests. Losses from insect infestations or disease may become disastrous at the individual farmer level; when extensive geographical areas are involved, the national economy may be seriously affected. Many of the striking reductions in the genetic diversity of our hybrid corn base were well established before 1960. It is possibly not accidental that all of our serious widespread corn epidemics have occurred since that date. These include: (1) The maize dwarf mosaic virus and corn stunt mycoplasma attacks of 1964; (2) *Phyllosticta* leaf blight in 1968; and (3) leaf blight in 1970.

The epidemic of maize dwarf mosaic virus was most severe in the Scioto and Ohio River bottom lands. It has been estimated that this disease caused a loss of $5 million in 1964 in Ohio alone. Screening of existing lines revealed some that have a satisfactory level of resistance. Susceptible lines were rapidly replaced, and resistant hybrids now

PLATE 3   Cornfield near Des Moines, Iowa, with a picker at work (Photo courtesy W. Brown, Pioneer Hi-Bred International, Inc.).

minimize losses throughout the area. The disease persists and, if susceptible hybrids are planted, heavy losses still recur.

Resistance to corn stunt, which occurs widely throughout the South, is relatively rare among the publicly developed lines. As a result, corn acreage and production have dropped drastically in some areas such as the Yazoo Delta. The best resistance found thus far has come from Latin American types of corn. In general, these are not satisfactory for commercial use, but resistance may be transferred to adapted types.

Yellow leaf blight (caused by *Phyllosticta maydis*) brought on a minor epidemic in the northern states in 1968. Differences in resistance among inbred lines have been observed, and hybrids with adequate resistance will rapidly replace the more susceptible types. *Kabatiella* (eyespot) caused damage in some areas in 1970. Whether or not this disease will persist and continue to cause losses cannot be predicted at this time.

GERM PLASM RESOURCES

Considerable effort has been expended on the collection and preservation of indigenous varieties of corn that, for a time, achieved some degree of commercial usefulness, and of the inbred lines developed from them. The Rockefeller Foundation, with the cooperation of the countries involved and with some financial assistance from the International Cooperation Administration (now the Agency for International Development [AID]) conducted an extensive program of collection, classification, and preservation of the races of corn from Latin America. Typical samples are currently in storage at the Federal Seed Storage Facility at Fort Collins, Colorado, in Mexico City, Mexico, and in Medellin, Colombia.

More recently, the original Latin American collections have been augmented and, with few exceptions, the maize of most areas of Latin America is now well represented in the seed banks in Mexico and Colombia.

Comprehensive collections have also been made in Italy, Spain, and Yugoslavia, and on an incidental basis in other parts of Europe. Most of these collections are in storage at Instituto Di Ricerche Orticole, Minoprio, Italy.

In the early 1950's, some effort was made to collect and preserve the open-pollinated varieties existing at that time in North America.

These materials, representing several hundred varieties, are maintained at the Plant Introduction Station, USDA, Ames, Iowa.

Unfortunately the availability of sizeable germ plasm resources does not necessarily ensure direct commercial usefulness. Exotic materials tend to be poorly adapted and therefore low yielding under conditions prevailing in the United States. In addition they usually exhibit other plant or ear characteristics that make them unacceptable to farmer or consumer.

Some exotic collections may be extremely valuable, however, in providing sources of resistance to various diseases or insects. When used simply as a source of inherited resistance they make no real addition to genetic diversity. They must be utilized as one feature of a wider program if they are to make a contribution to our present genetic base.

Ideally, breeding programs should be sufficiently broad in scope to further both immediate and long-range objectives. Until now, major reliance has been placed on short-term procedures and objectives, with primary emphasis on parental diversity and on the orderly development and distribution of superior hybrids. The short-term procedures include the choice of superior parental material (initially varieties, subsequently superior hybrids or synthetics) followed by the isolation and evaluation of inbred lines and the commercial utilization of the best performing combinations. This procedure starts with adapted material and through successive repetitions or cycles insures that each new generation of hybrids will represent an improved performance level. The genetic divergence between selected parents responsible for superior performance is enhanced through the system of test evaluation. The general practice of using good hybrids as source material for the development of new lines, however, insures a gradual reduction in the total genetic base.

BROADENING THE GENETIC BASE

In practice, the genetic base for corn can be effectively broadened only through the development and utilization of new genetic sources equal in performance to lines and hybrids now in commercial use. These new lines or gene pools may be developed from: (1) Indigenous material; (2) indigenous crossed with exotic; or (3) wholly exotic sources. Improvements in any one of these three groups will require long-term effort, the last solution requiring the longest time.

The direct sampling of open-pollinated varieties has been largely replaced by sampling from superior single- or double-cross combinations or through the improvement of existing lines by backcrossing. An alternative to the use of varieties as source populations has been the direct sampling of synthetics. Some of the first of these were developed by combining 16 to 24 lines in a common gene pool. The component lines were chosen because of certain characteristics, e.g. Stiff Stalk Synthetic, Low Ear Synthetic. Several valuable lines have been derived from synthetic varieties (B14, B37, N28; see Table 4).

Population improvement through recurrent selection techniques offers greater promise than does direct sampling. A considerable body of evidence is now available from quantitative genetic studies indicating that recurrent selection is an effective tool for improving both population performance and interpopulation hybrid vigor. Direct comparisons of the relative efficiency of different types of recurrent selection for population improvement are difficult because of differences in the base populations, in the type and intensity of selection practiced, and in the procedural details used in evaluation (Sprague, 1966). All methods have been shown to be reasonably effective. The gain in yield with successive cycles of selection has been much less striking than the corresponding gains in interpopulation hybrid vigor (Hallauer, 1972). The best of the current interpopulation hybrids have given yields approximately equal to standard double-cross hybrids.

If population improvement can be accomplished without a significant reduction in genetic variability, quantitative genetic theory predicts that the best double- or single-cross hybrids from a series of random lines from such selected populations would give additional yield increases of 20 to 30 per cent respectively. If the predictions on hybrid performance are realized, the development of new hybrids, equal or superior in yield and unrelated to those in current use, should pose no insurmountable problem.

Populations involving indigenous varieties crossed with exotic ones have been developed within the seed industry. Some of the public institutions have reoriented their corn programs, giving major emphasis to population improvement and using both indigenous and exotic materials. Lines from such populations are now in commercial use, and others will undoubtedly be used as developmental and evaluation studies progress.

Less work has been done with wholly exotic material. Direct inbreeding would be very inefficient; some type of population improvement must be used to increase the performance and acceptability level before commercial utilization becomes possible.

Cumulative experience indicates the importance of certain general procedures involved in the development and improvement of populations. Where several strains are to be combined to form a new gene pool (adapted, exotic or exotic X adapted), several generations of random mating should precede any artificial selection. Bulking of equal quantities of seed of the selected components is a wasteful and inefficient procedure. A half-pedigree procedure is preferable because it tends to conserve both cytoplasmic and genetic contributions. If the components are maintained as separate entities replicated within the isolated planting, then similarity and general appearance among entities will provide a rough measure of the approach to random mating. The general adoption of this new technology—recurrent selection and population improvement—should lead to a rapid broadening of our current genetic base.

Procedures for maintaining cytoplasmic diversity are less obvious. Techniques for measuring cytoplasmic differences are essentially nonexistent except for cytoplasmic male sterility, and certain chlorophyll and growth patterns. Until basic research in molecular cell physiology provides a basis for characterization and evaluation of possible differences, little progress can be expected in clarifying the role of cytoplasm.

Several cytoplasmic sterile strains were released as breeding stocks in 1970. These appear to fall into three distinct groups based on differential fertility restoration. They are coded as T, S, and C. The T source was first used in the early 1950's and was present in approximately 75–80 per cent of the corn acreage in 1970.

It led to the *Phyllosticta* and *Helmintrosporium* outbreaks. Lines with T cytoplasm have been identified, however, that have nuclear genes to provide an acceptable degree of resistance to one or both of these leaf diseases. Future use of this cytoplasm should be limited to lines having demonstrated resistance.

The S source of cytoplasmic sterility was utilized commercially in the 1950's but was discarded because of environmental instability. Some commercial use is being made of another cytoplasmic source (C), but its general utility remains to be demonstrated. The blight epidemic of 1970 has made all types of cytoplasmic male sterility somewhat suspect.

A return to reliance upon manual detasseling poses undesirable risks involving both labor supply and weather conditions, and intensive efforts should be directed toward the development of mechanical detasseling devices and the evaluation of gametocides.

SUMMARY

Plant breeding improvements are normally accompanied by a drop in genetic diversity. Such improvements must contribute direct economic benefits if they are to be commercially acceptable. The magnitude of the benefits achieved in corn is illustrated by the fact that we now produce over 50 per cent more corn on 25 per cent fewer acres than we did in the days of the open-pollinated varieties. These yield increases have been associated with a continuing succession of hybrid substitutions, changes in agricultural technology, and in methods of seed production. Each of these changes has involved a decrease in genetic diversity; dependence upon a decreasing number of inbred lines, and on a single cytoplasmic type (Texas male sterile).

Decreasing genetic diversity is accompanied by an increase in genetic vulnerability and an increased risk of economic loss caused by some new parasite, insect pest, or unusual environmental stress. If the hybrid affected is grown on a limited acreage, losses will be comparably limited. If the hybrid affected is grown extensively, the losses incurred are correspondingly increased.

The obvious course of action is to restore a necessary measure of genetic diversity through the development of new and unrelated types that are resistant to or tolerant of the new hazard. This will be difficult if we rely solely on breeding methods in general use. New breeding procedures have been developed that promise to conserve genetic diversity and at the same time provide continuing benefits as measured by significant increases in production.

REFERENCES

Duvick, D. N. 1959. The use of cytoplasmic male-sterility in hybrid seed production. Econ. Bot. 13:167–195.

Grogan, C. O., P. Sarvells, J. O. Sanford, and H. V. Jordan. 1965. Influence of cytoplasmic sterility on dry matter accumulation in maize (*Zea mays* L.). Crop Sci. 5:365–367.

Hallauer, A. R. 1972. Third phase in the evaluation of synthetic varieties of maize. Crop Sci. 12:16–18.

Noble, S. W., and W. A. Russell. 1963. Effects of male sterile cytoplasm and pollen fertility restorer genes on performance of hybrid corn. Crop.Sci. 3:92-96.

Richey, F. D. 1922. The experimental basis for the present status of corn breeding. J. Am. Soc. Agron. 14:1-17.

Russell, W. A., and F. Marquez-Sanchez. 1966. Effects of cytoplasmic male sterility and restorer genes on performance among different genotypes of corn (*Zea mays* L.). Crop Sci. 6:294-295.

Shull, G. H. 1908. The composition of a field of maize. Am. Breed. Assoc. Rep. 4:296-301.

Shull, G. H. 1909. A pure line method of corn breeding. Am. Breed. Assoc. Rep. 5:51-59.

Sprague, G. F. 1966. Quantitative genetics in plant improvement, p. 315 to 354. *In* K. J. Frey [ed.], Plant Breeding. Iowa State Univ. Press, Ames, Iowa.

Stringfield, G. H. 1964. Objectives in corn breeding. Adv. Agron. 16:102-137.

CHAPTER 9

Wheat

Contents

INTRODUCTION	120
CLASSIFICATION	121
ADAPTATION	122
DISTRIBUTION	123
GENETIC DIVERSITY	123
Germ Plasm Phenomena	123
Danger Areas	127
WHEAT BREEDING PROGRAMS	129
Germ Plasm Availability	129
Trends in Breeding Programs	133
VULNERABILITY OF WHEAT TO DISEASE	137
Host–Parasite Interaction	137
Specificity; Nonspecificity	
Environmental Factors	142
Unforeseen Vulnerability	143
Control Practices	144
VULNERABILITY OF WHEAT TO INSECTS	145
Conditions Influencing Vulnerability	146
Genetic Control of Insects	148
Summary	150
OUTLOOK	151
Strengths	151
Weaknesses	152
Research Needs	152

119

INTRODUCTION

Wheat has long been vulnerable to parasites, insects, and environmental hazards, resulting in frequent devastating losses. Early writings from the Mediterranean and Near East contain accounts of wheat losses from rust and insects. The great North American stem rust epidemic of 1904 destroyed a major part of the United States wheat crop. There has been a disturbing succession of subsequent stem, leaf, and stripe rust epidemics as well as severe losses from Hessian fly, aphids, the smuts, viruses, and numerous other diseases and insects.

Wheat is also vulnerable to its environment. Annual heavy losses are recorded from drought, hail, excessive summer heat, and low winter temperatures. One has only to recall the extended drought of the 1930's and reflect on that of 1971 to be impressed with this kind of vulnerability. When the farmer permits the fertility of his soil to become depleted, his wheat crop is substandard. This condition led Nobel prize winner Norman Borlaug (1968) to say recently, "The poor insects in unproductive wheat fields are as unfortunate as the man who is trying to grow the wheat."

120

Whereas this chapter is confined to a discussion of the vulnerability of wheat itself, most of the information and recommendations apply as well to such other cereal crops as oats, barley, and rye. The importance of these latter species in some regions of the world is well recognized.

CLASSIFICATION

Wheat belongs to the genus *Triticum,* the many species of which can be divided into three groups according to the number of chromosomes in their reproductive cells. Most of the world's cultivated wheat belongs to the 21-chromosome species usually called common wheat and known botanically as *Triticum aestivum* L. em. Thell. The other agriculturally important species, *Triticum durum* Desf., belongs to the 14-chromosome group. The seven-chromosome wheat (Einkorn), *Triticum monococcum* L., although grown by early man, is currently of little importance except in a few areas of the Near East.

Wheat also is classified according to its habit of growth. Winter wheat requires low temperatures and short days during its early stages to reproduce itself. Such wheat is seeded in the autumn or early winter of the year preceding harvest. Spring wheat requires no period of low temperatures and may be seeded in the early spring for summer harvest. Between the true winter and spring classes is an array of intermediate types that exhibit less pronounced temperature and photoperiodic requirements. The intermediate wheats are usually managed as winter wheat, but their production is confined mainly to the regions of moderate winter temperatures. Even spring wheats are managed this way in regions having mild winter climates.

Autumn-sown wheat occupies a substantially larger portion of the world's total wheat acreage than does spring wheat. In regions where both kinds are grown, winter wheat matures earlier and is usually more productive than spring wheat.

In the United States and in many other countries wheat also is classified according to grain color and texture. Wheat grain may be white or red, and its texture ranges from hard to very soft. The hard-textured common red wheats in the United States are used mainly for bread making, whereas the soft-textured red and white wheats are used for pastry, crackers, and sweet goods. The hard durum wheats are used chiefly for macaroni products.

FIGURE 1   Harvesting wheat on the Great Plains: one variety from horizon to horizon. (Photo courtesy V. Johnson, USDA.)

ADAPTATION

Wheat, adapted to the widest geographic range of any of the cereals, is grown throughout the world wherever conditions of temperature, moisture, and soil permit cultivation. Wheat harvest is in progress in every month of the year in some part of the globe. The crop can be grown in almost every kind of arable soil from sea level to 10,000 ft elevation, though the main wheat regions of the world lie between latitude 30° and 55° in the northern hemisphere and 25° and 40° in the southern hemisphere.

Annual precipitation in the northern and southern hemisphere wheat production belts ranges from 12 to 45 in.; more than one-half of the world's wheat area is in the 15- to 25-in. annual precipitation range, and 10 per cent occurs in areas where annual rainfall is less than 15 in. Only 15 percent of the world wheat acreage occurs where annual precipitation exceeds 35 in. In recent years, wheat irrigation in some countries has increased significantly. The spectacular "Green Revolution" in India, Pakistan, and the Near East is based on an intensive irrigated-production system. Similarly, the emergence of Mexico as a self-sufficient country in wheat production is associated with the development of areas of intensive irrigated production. However, the foundation of the world's wheat production continues to be dryland or rain-fed wheat.

Spring wheat is concentrated in regions where the winters are dry and cold with little snow cover. Spring wheat also is grown in areas with very mild winters where it is managed as a fall-sown crop. In the dry cold regions, fall-sown wheat cannot survive; and in regions of very mild winters its cold requirement for germination is not satisfied. Typical examples of such regions in the northern hemisphere are the northern plains of the United States and Canada, and Siberia, Kazakhstan, and Ural of the USSR, where winters are extremely harsh. Mexico, India, and Pakistan are typical of the mild-winter regions where spring wheat is seeded as a winter crop.

Winter wheat is favored over spring wheat in all areas in which fall seedings will survive, typically the central and southern plains of the United States, Western Europe, southern USSR, and the Balkan countries.

DISTRIBUTION

Wheat acreage and production exceed those of any other grain crop. It ranks next to rice in importance as a food crop and is the principal food for about one-third of the world's people. Table 1 shows world wheat acreage and crop size, 90 percent of which are concentrated in the northern hemisphere.

In the United States, 42 states reported wheat acreages in 1969, and wheat was a major crop in 20 states (Table 2).

GENETIC DIVERSITY

GERM PLASM PHENOMENA

The world's wheats are survivors of gene pools developed from cross breeding, mutations, natural selection, and other evolutionary forces. Wheat workers now recognize that preservation of the extensive genetic diversity of wheat may be threatened. Man affects the diversity and distribution of germ plasm through changes in land use on which wild wheat types grow and through the choice of varieties that he sows. In modern wheat culture the products of plant breeding are quickly exploited over wide areas. Ancient diverse varieties with long

TABLE 1 World Wheat Distribution in 1969

Continent or Country	Acreage (1000 acres)	Yield per Acre (bu)	Production (1,000 metric tons)	New Varieties Made Available[a] 1962–70
North America (Canada, Mexico, United States)	74,389	29.8	60,364	127
South America (Argentina, Brazil, Chile, Uruguay)	18,837	19.5	9,975	45
Western Europe (includes 16 countries)	42,755	38.9	45,248	198
Eastern Europe (Bulgaria, Czechoslavakia, East Germany, Hungary, Poland, Romania, Yugoslavia)	26,588	35.7	25,410	53
USSR	163,827	14.6	76,600	7
Asia (Iran, Iraq, Syria, Turkey, China, India, Japan, Korea, Pakistan)	159,791	15.4	66,867	56
Africa (Algeria, UAR Morocco, Tunisia, Repub. of South Africa)	19,671	13.2	7,089	42
Oceania (Australia, New Zealand)	24,474	17.0	11,322	14
World Total	530,332	20.2	291,866 (10,726,075,500 bu)	

A metric ton = 36.75 bu

[a] Information from Crops Research Division, ARS 34-117, 1970.

Source: USDA Agricultural Statistics, 1970.

evolutionary histories may be replaced by those having germ plasm of limited diversity. Such changes can occur almost "overnight" as was the case in the highly successful introduction of Mexico-derived varieties to the Mideast and elsewhere.

Modern wheat breeding has been strongly influenced by the pureline method. New varieties sown by farmers are largely single genotypes, whereas the varieties that they replaced often comprised several genotypes with considerable diversity. Loss of these old "land varieties" further restricts the genetic diversity of wheat and increases its vulnerability to various hazards.

TABLE 2   Seeded Acres of Wheat in the 17 Major Wheat-Producing States of the United States in 1969

State	Total Seeded Acres (1,000 acres)	Winter Wheat (1,000 acres)	Spring Wheat (1,000 acres)	Durum Wheat (1,000 acres)
Texas	4,124	4,124	–	–
Oklahoma	5,299	5,299	–	–
Kansas	10,767	10,767	–	–
Nebraska	2,999	2,999	–	–
Colorado	3,115	3,075	40	–
South Dakota	2,184	740	1,200	244
Montana	3,809	2,445	1,129	235
North Dakota	6,938	110	3,997	2,831
Minnesota	852	22	740	90
Total	40,087	29,581	7,106	3,400
Missouri	1,190	1,190	–	–
Illinois	1,359	1,359	–	–
Ohio	1,105	1,105	–	–
Indiana	956	956	–	–
Michigan	679	679	–	–
Total	5,289	5,289	–	–
Washington	2,890	2,590	300	–
Idaho	1,166	929	237	–
Oregon	834	774	60	–
Total	4,890	4,293	597	–
17-State Total	50,266	39,163	7,703	3,400
United States	54,312	43,120	7,786	3,406

Source: USDA Agricultural Statistics, 1970.

Because of its genetic makeup, wheat has the potential for much genetic diversity. There are 14, 28 and 42 chromosome types. The near relatives of wheat (*Aegilops, Agropyron, Secale, Haynaldia*) also possess 14 chromosomes or a multiple of 14 and many hybridize readily with wheat.

Geneticists have recognized that successful crops (such as wheat) with more than one set of chromosomes possess a great store of potential genetic variability. Thus wheat breeders have been able to develop a wide array of genotypes useful for improving adaptation and overcoming production hazards. Genetic evidence on the evolution of wheat came in this century when three natural groups based on chromosome number were recognized; thus wheat is a polyploid crop. The important bread and durum wheats arose from seven-chromosome wheat (Einkorn) and two *Aegilops* species by hybridization followed by chromosome doubling.

A co-mingling of the three species led to rapid increase in genetic diversity and variability. Continued natural and man-imposed selection over a long period transformed the raw hybrids into functional types that we now cultivate. Flow of genetic material among the various wheat species served as an effective device for rapid buildup of variation and the assurance of genetic flexibility. The apparent evolutionary success of the wheat group probably resulted from its ability to absorb alien genetic material or "to behave as genetic sponges." Future search for genetic variability in wheat should include centers of origin and diversity. Both cultivated and noncultivated types should be collected.

Large numbers of new types of wheat have been artificially synthesized (Quisenberry and Reitz, 1967). Species of *Triticum* can be hybridized with species of the other four genera of the subtribe *Triticinae*. At least 18 different intergeneric and 59 interspecific combinations have been described. Some of these intergeneric hybrids have been named *Triticale* (*Triticum* X *Secale*); *Agrotricum* (*Triticum* X *Agropyron*); and *Haynaltricum* (*Triticum* X *Haynaldia*). These artificially evolved species show promise of adding new levels of diversity.

Wheat chromosomes are homologous. Cultivated bread wheat has three sets of seven chromosomes each. For each chromosome of a set, there is a chromosome in each of the other two sets that is closely related in structure, function, and genetic information. Undoubtedly, numerous genes on these related chromosomes regulate similar genetic events, which is why the inheritance of economic traits in wheat is often complex. Also, this situation makes possible the accumulation and preservation of variability through mutations (natural or induced). The repetition of genetic information throughout the sets of chromosomes of wheat and its near relatives has served as a buffer system that has allowed an enormous number of natural and induced mutations to occur, survive, and add to the genetic diversity of wheat.

A breeder's own material is frequently his most useful source of genetic diversity. Every breeder has a working collection of varieties and breeding stock. Most of these are transitory, of changing composition and of short duration. Unfortunately, such stocks have limited value, since they are frequently poorly preserved and restricted in their dissemination. Currently steps are being taken for preservation and dissemination of such germ plasm on an international basis.

Mutations are another source of genetic diversity in wheat. They occur naturally or can be induced. Recent advances in mutation breeding suggests this method can be used to create new genetic variability when the need arises. To tailor-make genetic diversity of wheat would be a scientific breakthrough. Mutation induction techniques also have been used successfully to transfer resistance to diseases from *Aegilops, Agropyron,* and *Secale* to wheat.

Although wheat is a self-pollinating crop, cross-fertilization occurs in nature at sufficient rates to allow for recombination of the genetic variability found in the crop. This process is by no means static. The reshuffling of genetic information of wheat is continuous and can be usefully guided by man. Unfortunately, few wheat workers employ breeding schemes designed to encourage large-scale genetic recombination, a situation that must be remedied.

DANGER AREAS

The fall in popularity of wheat varieties as the result of disease and insect epidemics demonstrates the need for a broad base of germ plasm for disease and insect resistance. In recent years the life expectancy of wheat varieties in the Pacific Northwest has been about five years. The variety Elgin was grown for six years before common bunt became destructive to it. Elmar replaced Elgin, but within five years it was compromised by another bunt race. Omar replaced Elmar in 1957, but it had to be replaced by Gaines in 1961 because of extreme stripe rust susceptibility. Gaines and Nugaines probably will be replaced by two new varieties more resistant to foot rot and stripe rust.

Monoculture of wheat over a wide area probably represents the chief potential vulnerability. Theoretically, a microorganism, insect, or condition capable of injuring one plant could similarly affect all other plants of the variety. Thus, the concentrated areas of Bezostaia in the USSR and Eastern Europe, Scout in the central plains of the United States, and Gaines in the Pacific Northwest are potential trouble spots.

The widespread acceptance and use of semidwarf wheats is similarly dangerous. Several factors are involved. The yield potential of these wheats can be realized only if they are managed properly and remain healthy. There must be optimum stands with heavy use of fertilizers and water to assure luxuriant plant growth. Under such management conditions, an ideal environment for foliar diseases has been created. The effect is accentuated in semidwarf varieties. For example, when resistance to stripe rust is lacking among semidwarf varieties, losses in excess of 60 percent may occur. Comparative losses among susceptible nondwarf sibling lines are about 30 percent. The microenvironment of semidwarfs together with intensive management practices are conducive to heavy losses from many foliar diseases.

The management systems required for the full exploitation of the potential of semidwarf varieties pose an indirect threat to them. With the intensification of wheat culture, especially at higher levels of fertilization, more rather than less trouble from diseases and insects is likely to be encountered. Some excellent examples of such "sleeping threats" to Gaines are the rapid and large increases in foot and root rot diseases.

As yet, no major difficulty has been found associated with the day-length insensitivity trait of the semidwarf wheats except vulnerability to late spring frosts in the northern regions. Preliminary evidence also suggests that derivatives of these light-insensitive wheats can be damaged severely by the straw-breaker disease.

It is too early to say whether hybridization will directly or indirectly increase the genetic vulnerability of wheat. An obvious analogy can be drawn between hybrid corn and hybrid wheat. Most wheat workers have concentrated on a single cytoplasmic sterile type, *Triticum timopheevi*, just as corn workers placed most of their emphasis on the Texas T cytoplasm type that ultimately led to the southern corn leaf blight epidemic. The wheat sterile cytoplasm base is equally as narrow and could be a dangerous weakness. It is already known that ergot will be as serious a problem in hybrid seed production in some areas, as it is in sorghum. Similarly, there could be dangerous increases in loose smut and Karnal smut in hybrid seed production fields. *Fusarium* head blight or scab may also pose a threat to hybrid wheat, since the open-flower nature of the female lines would encourage infection.

Widespread use of certain important genetic traits could pose a threat to stable wheat production. Examples of genes now being utilized extensively and often in similar genetic backgrounds are the semidwarf genes of Norin 10, the high protein gene of Atlas 66, the light-insensitive genes of the CIMMYT* wheats, the compactum gene of Omar derivatives, and the restorer genes of *Triticum timopheevi.*

*Centro Internacional de Mejoramiento de Maíz y Trigo (International Maize and Wheat Improvement Center), Mexico City.

When vulnerabilities are linked to such genes, there is serious potential threat to wheat production.

With the advent of the cyto-sterile restorer system and a good genetic sterile system, wheat breeders have techniques for promoting recombination of characters and perpetuation of genetic diversity. Since the Variety Protection Act stresses novelty and uniformity, there is an implied need for genetic homogeneity to qualify for patent protection. Indeed, this law may discourage wheat breeders from pursuing the "genetic diversity concept" of plant pest biological control. If so, it poses a danger.

WHEAT BREEDING PROGRAMS

GERM PLASM AVAILABILITY

Wheat varieties are developed and released for farmer use much more quickly now than in the past. The trend is distinctly toward increasingly larger numbers of varieties from which the farmer can choose. Concurrently there is a more rapid farmer acceptance and use of good new varieties, which has led to near monoculture in some areas of the world. This tendency has been accelerated by the demands of processors, exporters, and consumers for a uniform product.

Grazing, sometimes overgrazing, diminishes or eliminates species from their native range. Tillage of new land areas and use of land for nonagricultural pursuits further reduces the native reservoirs of wild *Triticums* and relatives.

Wheat breeding has resulted in the creation of highly efficient, productive, and pest-resistant varieties. It is unthinkable that we would deliberately revert to less-productive types of varieties merely for the sake of preserving germ plasm in farmers' seed. However, the multiline approach to variety development is one means to increase diversity without loss of productivity.

A substantial amount of variability of wheat already may have been lost. When Nikolai Vavilov did his classic studies on wheat, his collection numbered over 30,000 entries. Most of this material no longer exists. The USDA wheat collection consists of over 19,500 accessions. Most of the present collection, which was contributed by 74 countries, has been assembled since 1948 because many of the original stocks were lost.

Reitz and Craddock (1969) point out that the USDA collection tends to be fortuitous in that seed was obtained from areas accessible to collectors rather than areas of known diversity. They suggest that in the search for disease and insect resistance, the effort was not concentrated in those areas where evolutionary pressure was known to be operating. There is a clear need to obtain seed from the Communist-bloc countries, particularly USSR and China.

Recently the Food and Agriculture Organization (FAO) and International Atomic Energy Agency (IAEA) jointly have begun to compile an international list of researchers holding stocks of wheat and related species and genera. Stocks from 27 countries and also held in the USDA wheat collection are summarized in Table 3. The degree of duplication and composition of the collections in other countries are essentially unknown. The major "emergency" areas

TABLE 3   Stocks in the USDA Wheat Collection, from 27 countries, based on a 1970 FAO Survey

Country	In USDA Collection	Held in Country according to FAO Survey
Afghanistan	948	—
Algeria	57	—
China (Mainland)	1,315	—
Egypt	286	—
Ethiopia	1,127	780
Greece	158	2,000
India	1,097	8,272
Iran	652	10,242
Iraq	83	—
Israel	58	2,620
Italy and Sicily	379	9,000
Jordan	59	—
Korea	47	—
Lebanon	20	—
Libya	2	—
Mongolian Repub.	0	—
Morocco	69	—
Nepal	0	80
Pakistan	187	7,092
Sikkim	0	—
South Africa	214	—
Sudan	1	—
Switzerland	950	5,140
Syria	64	30
Tunisia	198	1,000
Turkey	2,493	8,100
USSR	1,313	21,164

(where germ plasm may be lost) are Asia, parts of Africa, Southern Europe, and countries bordering the Mediterranean. Wheat of greatest antiquity probably exists there, and the accelerated rate at which new varieties are replacing old ones constitutes a serious threat if valuable germ plasm is to be preserved. (Wheat is new to the Americas and Oceania, and their varieties are largely derived from older areas.)

The USDA Wheat Gene Bank is an additional important source of germ plasm. This is comprised of bulk seed from F_1 and F_2 plants donated by wheat breeders.

The problem of germ plasm availability is not entirely one of numbers. For instance, in the USDA wheat collection, many specific genes are known to be repeated in numerous varieties. The probability of adding altogether new genes may be diminishing unless exotic new gene pools are found and incorporated into the existing collection.

Is the ratio of germ plasm in world wheats in the wild and in farmers' seed to conserved stocks dropping at all, or dropping in some places at a catastrophic rate? Creech and Reitz (1971) think so. An FAO Panel of Experts of Plant Exploration and Introduction (1967) thinks so.

There is little doubt that the availability of natural variability in wheat is eroding rapidly. Agricultural progress in the Near East threatens to destroy natural wheat types traditionally known for their evolutionary history. Both wild and cultivated forms are in danger. Extensive collecting in these areas is imperative before it is too late.

Induced mutations are an additional means of adding to the collection. Reitz and Craddock (1969) have challenged mutation research to come of age and begin to develop important types that do not exist in nature. Leaders in plant mutation genetics have accepted the challenge and indeed are beginning to concentrate on creating new genetic diversity. The wheat mutation program at Washington State University has placed major emphasis on developing unique semi-dwarf types, dominant dwarf lines, general resistance to stripe rust, restorer genes for cytoplasmic sterility and tolerance to *Cercosporella* foot rot.

In a recent symposium (1970), Brock suggested that induced mutations are a logical alternative to naturally occurring variation as a source of germ plasm for plant improvement. Evidence was presented that the frequency of induced mutations of small effects are roughly 10 times that of recessive lethals. This constitutes a tremendous store of genetic variability with obvious evolutionary consequences.

Dynamic improvements are possible through transgressive segregation when forms appear that surpass either parent. A classic example was the hybridization of the Japanese wheat Norin 10 and the Washington wheat Brevor. This cross gave progeny with outstanding yield potential. We now know that genes for yield came from both parents and resulted in offspring with yields unparalleled in existing germ plasm.

Similar cases can be cited for disease and insect resistance. Workers have reported transgressive inheritance for resistance to several wheat diseases and insects. Examples of this form of resistance have been shown for foot rot, snow-mold, stripe rust, stem rust, flag smut, and cereal leaf beetle. Breeding schemes that exploit natural levels of outcrossing to increase gene recombination and release genetic variability definitely have a place in perpetuating and increasing our germ plasm base.

An important source of unique germ plasm for wheat breeders has come from gene transfer from alien chromosomes to wheat chromosomes. This has been particularly true for disease resistance. To be of economic importance little deleterious effect must be associated with the exchange of alien chromatin with wheat chromatin. Alien transfer material with useful resistance includes leaf rust resistance from *Aegilops umbellulata,* stem rust resistance from *Agropyron elongatum,* stripe resistance from *Aegilops bicornis,* stripe and leaf rust resistance from *Agropyron intermedium* and leaf rust and powdery mildew resistance from *Secàle.* Production of alien transfer germ plasm is laborious. Evidence that parasites may be able to circumvent this form of resistance as readily as forms that already occur in wheat has tended to reduce enthusiasm for the system. Nonetheless, several of the above examples have maintained their wide protection against numerous races of the disease involved. For this reason the method has practical value and merits increased attention.

Collecting and preservation of wheat genetic stocks are only part of the problem. Basic documentation of wheat stocks represents an important phase of germ plasm use. There is greater usefulness and the genetic resources increase in value as documentation can be supplemented by descriptive information, evaluation data and records of identified genes. Since 1950 International Rust Nursery data have been recorded and disseminated as a regular part of the USDA-sponsored program. Through the combined efforts of the FAO and IAEA a program is being formulated aimed at the development of an international genetic resources information system. One function of this group is standardization of international recording of crop research data.

The need for uniform recording systems of wheat research data is obvious. Encouraging advances have recently been made on standardization of disease response code scales on an international basis. Agreement has apparently been reached on a 0 to 9 scale for recording data. Loegering (1968) suggested the logical use of a "guide code," derived from the actual disease data, to rate disease response. This is, in effect, a guide to response and lends itself to international use. A computer-based information retrieval system may be the ultimate in documentation of genetic stocks.

Painstaking effort has been spent on collection of information on many traits of various wheat collections, yet for the most part, this information stays hidden in inaccessible records of workers and experiment stations. It is indeed unfortunate that germ plasm of some of the most aggressive wheat improvement programs in the world is poorly preserved and undocumented.

Regular donations by all, not just a few, breeders must be made to the Wheat Germ Bank. Every year valuable first and second-generation seed stocks are unwittingly destroyed when they could be adding to our overall genetic diversity. Reitz and Craddock (1969) believe that breeders' own collections are frequently the best place in which to search for needed genes.

TRENDS IN BREEDING PROGRAMS

The current trends in wheat breeding provide clues to areas of potential vulnerability of wheat. Of the 263 varieties grown in the United States in 1969, varietal groups, Scout and Triumph, made up 25 percent of the total (Table 4). Nine varieties, five of which were hard red winters, made up 50 percent, and 18 varieties constituted 70 percent of the United States acreage. Of these, 10 were hard red winter wheats grown mainly in the southern and central Plains states, five were common hard red spring and durum wheats grown in the northern Plains states, one was a soft red winter wheat grown in the upper Midwest, and two were soft white wheats grown in the Pacific Northwest. Since 1962, 93 new wheat varieties have been developed and released for commercial production in the United States, eight of which are currently among the 20 most widely grown varieties. One

TABLE 4 Acreages and Areas of Production of the Major Wheat Varieties in the United States in 1969

Variety	Market Class	Acres Seeded	Primary Area of Production	Percent of Total U.S. Acreage	Cumulative Percent
Scout	Hard red winter	7,789,820	N. Mex., Texas, Okla., Kans., Colo., Nebr., S. Dak.	14.4	
Triumph	Hard red winter	5,971,943	N. Mex., Texas, Okla., Kans., Mo.	11.0	25.4
Monon	Soft red winter	2,352,597	Ind., Ohio, Ill., Mo., Tenn., Ark., Kentucky, Mich.	4.3	
Wichita	Hard red winter	2,247,108	N. Mex., Okla., Texas, Wyoming, Kans., Colo.	4.1	
Leeds	Durum	2,115,977	N. Dak., S. Dak., Mont., Minn.	3.9	
Lancer	Hard red winter	2,005,667	Mont., S. Dak., Wyo., Nebr., Kans., Colo.	3.7	
Chris	Hard red spring	1,945,805	N. Dak., Minn., S. Dak.	3.6	
Warrior	Hard red winter	1,519,579	Mont., Colo., Wyo., Nebr., Kans	2.8	
Nugaines	Soft white winter	1,454,145	Wash., Idaho, Ore., Utah	2.7	50.3
Wells	Durum	1,363,350	N. Dak., Mont., S. Dak., Minn.	2.5	
Fortuna	Hard red spring	1,359,072	N. Dak., Mont., S. Dak.	2.5	
Gage	Hard red winter	1,333,366	Nebr., Kans., Mo., Ill., Wisc., S. Dak.	2.5	
Kaw	Hard red winter	1,323,770	Kans., Okla., Texas, Colo.	2.4	
Winalta	Hard red winter	1,136,245	Montana	2.1	
Cheyenne	Hard red winter	1,097,098	Mont., Wyo., Nebr., Colo., N. Mex.	2.0	
Gaines	Soft white winter	1,043,235	Wash., Idaho, Ore., Utah	1.9	
Manitou	Hard red spring	1,024,346	N. Dak., Minn., S. Dak.	1.9	
Bison	Hard red winter	1,015,260	Kans., Colo.	1.9	70.2

Source: Distribution of varieties and classes of wheat in the United States in 1969. Agricultural Handbook (in press).

134

of these, Scout, released in 1963, occupied 14.4 percent of the United States acreage in 1969 and is now the most widely grown variety in the country.

Although a large number of new wheat varieties have been made available to farmers throughout the world, the rapid acceptance and popularity of a few varieties has led to near monoculture in some important wheat-producing regions. This highly vulnerable situation exists in the United States, where a relatively small number of varieties dominates the wheat acreage. The acreage percentages in 1969 occupied by a few of the most popular varieties in some states appear in Table 5.

In Canada six spring wheat varieties accounted for 86 percent of the wheat acreage in 1969; Manitou alone constituted 46 percent of the acreage. The wheats of Mexico are largely semidwarf spring varieties (chiefly INIA 66) with production concentrated in the irrigated valleys of Sonora, Sinaloa, and Baja California.

A Russian winter variety, Bezostaia, dominates the wheat acreage of Eastern Europe and southern Russia; it is probably grown on more

TABLE 5   Wheat Varieties as a Function of Acreage

Variety	State	Percent of wheat acreage
Wakeland	Florida	98.7
Blueboy	North Carolina	78.6
	Virginia	56.7
	South Carolina	38.2
Redcoat	Pennsylvania	74.2
	Maryland	65.4
	New Jersey	61.4
Genesee	Michigan	62.0
Monon	Indiana	55.1
	Ohio	49.2
	Tennessee	48.2
	Illinois	37.6
Nugaines	Washington	52.7
	Oregon	51.7
	Idaho	38.6
Chris	Minnesota	50.4
Scout	Kansas	42.2

than 40 million acres. A second Russian winter variety, Mironovskaya 808, is more winter hardy than Bezostaia and may be grown in the USSR on an acreage larger than that of Bezostaia.

Spectacular increases in wheat production in India, Pakistan, Iran, Afghanistan, and Turkey have been achieved since 1966, associated with the introduction of high-yielding semidwarf wheats from Mexico. Acreages of these high-yielding varieties in 1970 are shown in Table 6.

The extensive and world-renowned results of the CIMMYT (Mexico) program of wheat production research pioneered the "Green Revolution." Three ingredients that fostered this accomplishment were disease resistance, semidwarf growth habit, and day-length insensitivity. Although not universal by any means, many wheat breeding programs now are placing major emphasis on the use of the semidwarf growth habit. Such short wheat received worldwide recognition through the feats of Norman Borlaug.

The semidwarf trait can be traced to the single Norin 10/Brevor type developed by Orville Vogel. Wheats with this trait now are grown on 26 million acres in Asia, Europe, Africa, North America, and South America, though there is potential danger in having a single semidwarf genetic type gain such widespread popularity. While these varieties have a relatively wide genetic base for resistance to the three rusts, they are known to be vulnerable to such other hazards as leaf blights, root rots, insects, and poor seedling vigor.

Most wheat varieties are day-length sensitive. The trait has tended to restrict the use of individual varieties to relatively small production areas of similar day-length. The incorporation of genes for day-length insensitivity in varieties has significantly expanded the areas in which they could be effectively grown. The gene base for day-length

TABLE 6   Acreage of High Yielding Semidwarf Varieties in Asia, 1970

South Asia	Acres	Est. Percent of Total Acreage
South Asia		
India	15,100,000	38
West Pakistan	7,000,000	45
Afghanistan	360,000	6
Nepal	186,000	—
West Asia		
Turkey	1,540,000	8
Iran	222,000	2
Lebanon	4,200	3
Total	24,412,200	

insensitivity is narrow. There are known vulnerabilities associated with the trait, such as foot rot and damage from frost.

There has been an intensive effort since about 1960 toward practical utilization of hybrid vigor in wheat by using pollen-sterile cytoplasm derived mainly from *Triticum timopheevi*. The effort to date indicates that hybrid wheat is feasible and that economic levels of hybrid vigor can be attained. This may represent potential vulnerability since a single source of cytoplasmic sterility is involved. The danger may be overcome by taking advantage of the recent discovery of an apparently suitable chemical (to produce) sterility in male wheat. If highly encouraging early results are supported by more extensive large-scale field tests, wheat breeders will be able quickly to develop wheat hybrids free of many of the drawbacks of the cytosterile/restorer system.

There is a definite trend among wheat breeders to develop and screen large numbers of early generation materials. The tendency among breeders to adhere to pure line breeding is much less pronounced than in the past. Many programs now utilize a modified bulk system such as the second generation progeny test method. The result often means more genetic variability in new varieties.

The impact in the United States of the recently enacted Plant Variety Protection Act (Public Law 91–577) on trends in wheat breeding cannot yet be measured, although some forecasts seem possible. A major aspect of protection under the new law is that a new variety or stock must be novel to existing protected stocks. To achieve novelty, uniformity, and reproducibility, breeders may concentrate heavily on incorporation of easily identifiable morphological traits into new materials and on achieving a high degree of homogeneity in them. Some breeders are concerned that undue emphasis may be given to unimportant marking traits at the expense of performance, and that the broad germ plasm base and genetic diversity may be lost.

VULNERABILITY OF WHEAT TO DISEASE

HOST–PARASITE INTERACTION

In the monograph *Wheat and Wheat Improvement* (Quisenberry and Reitz, 1967) 49 diseases of wheat are described and discussed. These include 3 rusts, 4 smuts, 20 viruses, 6 root and foot rots, 3 spike and head blights, 6 foliage diseases, and 3 diseases caused by bacteria and

4 by nematodes. These 49 diseases of the wheat host are caused by approximately 60 different parasitic organisms. Each association of these organisms with wheat is controlled by the interaction of genes in the parasites and in the wheat plant. In the past, most studies have concentrated on the host and parasite as separate entities; thus very little is known about the interacting genetic systems. We have shied away from the hard task of studying these systems and have instead compiled a large literature of hypotheses, simple deductions, educated guesses, and poorly supported opinions. As a result we have the frightening probability that wheat is genetically vulnerable to damage by disease.

Specificity The experiences of the past half century have taught us that wheat is constantly vulnerable to yield reduction as a result of genetic changes in its parasites. The Canadian plant pathologist T. Johnson (1961) has reviewed this idea and referred to our plant breeding programs as "man-guided evolution" of parasites. He was speaking mainly of the fungi that cause the rust diseases of wheat. In the case of leaf and stem rust we have some knowledge of the nature of the genetics of the host-parasite association and have found a gene-for-gene relationship, called *specificity*. Studies on the genetics of specificity have demonstrated that for each gene for reaction in the host there is a corresponding gene for virulence in the parasite. For each of these there are alternative forms of the gene for either high or low reaction or pathogenicity. Low infection type (resistance) occurs only when the corresponding forms of the gene and parasite are for low reaction and low pathogenicity. High infection type (susceptibility) occurs if either host or parasite has corresponding forms of the gene for high reaction or pathogenicity. Thus "resistance" is genetically determined by both host and parasite, whereas "susceptibility" is genetically determined by either host or parasite. As customarily used, "resistance" and "susceptibility" are not strictly contrasting concepts.

In conventional breeding programs in past years, a gene for resistance corresponding to a gene for avirulence in the parasite was incorporated into a new variety of wheat. This placed tremendous selection pressure on the parasite fungus so that a single genetic variation in the fungus that made it better able to survive was quickly propagated, the new variety was damaged, and we identified a new race of the parasite. This is what Johnson meant by "man-guided evolution."

In one study (Rowell, Loegering, and Powers, 1963), irradiation of urediospores of the wheat stem rust fungus induced a single gene for pathogenicity to mutate from low to high pathogenicity at a rate of 1 in 1100 infections. Similar rates of mutation have been found in fungal parasites of other crops. Under natural conditions the mutation rate in the wheat stem rust fungus for a single gene would be far less—perhaps in the range of one in a billion infections. However, the number of infections that might occur in the wheat fields of the United States in a single year is in the billions of billions. Thus the opportunity for genetic change is almost unlimited. The effectiveness of this natural capability for mutation may be reduced in several ways:

1. If the wheat variety has two genes for resistance and the parasite has the two corresponding genes for avirulence then the parasite must acquire two mutations in the same spore to be virulent.

2. Usually genes for virulence are recessive. Thus a mutation to virulence in an individual that has both corresponding genes in the form for avirulence, i.e., homozygous for avirulence, will not be expressed until the virulence becomes homozygous through one of several mechanisms.

3. When a spore is produced that has the necessary genes for virulence, it must be deposited on a host plant.

4. Infection and development proceed only if the environment is favorable.

It is evident that, although mutation rates may be relatively high, the chances for a given mutation to survive are small. If it does survive initially, then lack of competition from other genotypes results in rapid build-up of the new genotype. In wheat mildew, where the parasitic part of the fungal life cycle is haploid, a mutation to virulence is immediately expressed. The more common development in this fungus of a new race with a new gene for virulence is perhaps due to this single factor.

The history of wheat breeding to prevent damage from stem rust has been the repeated development of new varieties carrying a "new" gene for resistance and the subsequent development of a race of the stem rust fungus with the corresponding gene for high virulence. To a large extent scientists have been successful over the past 40 years in keeping ahead of these changes in the parasite. A typical illustration would be for the durum wheat varieties grown in the upper Midwest. The first successful durum variety was Mindum, but the rapid increase of Races 17 and 38 of the fungus resulted in losses. Steward and similar durum varieties were released and grown successfully until the destructive epidemics of Race 15B in the early 1950's. Langdon durum served for a few years until a new strain of Race 15B became widespread. The Wells variety is the latest in this succession and continues to avoid damage from stem rust although one isolate of the stem rust fungus has been found that is virulent on it. A similar history could be detailed for other types of wheat and for other regions of the world. This potential for change in the population of the stem rust fungus provides serious genetic vulnerability in wheat. This is true not only for stem rust but for some other diseases of wheat as well.

Not only do genetic changes in the parasite represent a hazard, but unintentional changes in the host may also occur in breeding programs. Prior to 1961 stripe rust of wheat occurred in the Pacific Northwest but was considered to be of limited economic importance. By 1961 the older varieties of wheat had been largely replaced by new varieties. In the development of these new varieties some of the genes for resistance of the older varieties had been replaced by genes for susceptibility. As a result, in 1961 and following years, the stripe rust fungus became economically important even though no evident genetic change had occurred in the fungus. As with the rusts, the smuts of wheat have caused extensive damage to varieties either because in breeding programs genes for resistance were replaced by genes for susceptibility or a natural genetic change occurred in the pathogen.

Nonspecificity   Knowledge regarding the 41 diseases of wheat other than the rusts, smuts, and mildew is limited and thus the extent of the genetic vulnerability that is due either to the replacement of genes for resistance by genes for susceptibility in the host or to mutations from avirulence to virulence in the parasite is largely unknown. However, there is evidently another biological mechanism that limits the potential impact of these diseases on yield losses. Records of this phenomenon go back hundreds of years, but it was first studied intensively in the 1920's. Since that time the phenomenon has acquired many names, which have been collected by Thurston (1971) as follows: field resistance, tolerance, partial resistance, horizontal resistance, nonspecific resistance, nonspecificity, nonhypersensitive resistance, relative resistance, nonrace specific resistance, generalized resistance, mutigenic resistance, polygenic resistance, minor gene resistance, multigenic resistance, multiallele resistance, quantatively inherited resistance, and uniform resistance. To Thurston's list of terms may be added functional and morphological resistance. The phenomenon is referred to here as nonspecificity and can be distinguished from specificity (represented by the gene-for-gene relationship). Nonspecificity can be defined as any reduced amount or effect of disease that is not known to be specific. This definition is only an admission of lack of knowledge and by no means implies that the concept of nonspecificity is in error. It may well be that nonspecificity is a genetic character of the host alone, whereas we know that specificity is genetically determined by both host and pathogen.

H. C. Young, Jr. (1970) has called attention to the value of nonspecificity with respect to leaf rust of wheat. He says

> Although this factor (nonspecificity) was not specifically exploited, a satisfactory crop of wheat can be produced throughout the central plains area of the United States at the present time even in the presence of considerable amounts of leaf rust infection. It is also known, however, that the wheat varieties grown in that area 25 to 30 years ago did not have that capability, and when those same varieties are grown today in the presence of leaf rust their yields are very poor. Similarly, when wheat varieties from other parts of the world are introduced into the plains area of the United States many of them are actually killed by leaf rust adjacent to rows of the current commercial varieties which are producing satisfactory grain.

Workers around the world have demonstrated that certain varieties of wheat are less damaged by leaf and glume blotch than are other varieties. This observation perhaps is another illustration of nonspecificity since specificity has not been demonstrated in these diseases.

Nonspecificity in wheat could be extremely useful in breeding programs; however, it is much like the weather—everyone talks about it but few try to do anything about it. At present nonspecificity is poorly understood and not used in an orderly manner in breeding programs primarily because we do not know how to select for it, we have no knowledge of the genetic mechanisms that control it, and worse, we do not know how effective it would be in reducing losses caused by many of the diseases of wheat. Since nonspecificity must be under genetic direction, we do not know how vulnerable such protection would be to genetic change in pathogens. The example of leaf rust would indicate that nonspecificity could be very effective and that it could be quite enduring for maintaining high production of wheat.

ENVIRONMENTAL FACTORS

Disease, which results from the association of a host and a parasite, and the spread of plant disease are affected by environment (used here in the meterological and biological sense). Stem rust of wheat is an example of the complexity of the interrelationships of host-pathogen-environment. Urediospores of the stem rust fungus germinate only in the presence of free water. The growth of the fungus, prior to infection, is favored by temperatures of 65–75 F and by darkness. Infection, however, proceeds best at temperatures of about 85 F and in bright light, because the plant removes the carbon dioxide from around the infection site under these conditions. After infection has occurred, a period of warm, bright weather favors the rapid development of the disease. Following infection, there are some cases where the presence of a corresponding gene pair (*Sr6* for example) of host and parasite will restrict development of disease at 70 F or lower but not at 75 F or above. This kind of "resistance" is useful only where the average field temperatures are relatively low.

In such diseases as root and foot rots, leaf and glume blotch, and scab, the development and spread of the disease are strongly limited by the environment. Some of these many diseases cause yield reductions only under drought conditions; others only under cool, humid conditions; and still others only under hot, humid conditions. Thus there usually is a set of conditions ideal for development of each disease, conditions that may occur rarely, in which case damage from a given disease is seldom observed, but damage may be very severe in local areas almost every year. For example, wheat streak mosaic in some years has caused considerable damage to wheat. The virus is transmitted from plant to plant by a mite and the disease becomes important only under conditions that favor rapid reproduction and spread of viruliferous mites. The widely grown Mexican wheats have some diversity in genotype for reaction to some diseases but little diversity with respect to the semidwarf trait. This trait has made these wheats more vulnerable to such diseases as leaf blotch.

Combinations of disease often are more damaging than each disease alone. Leaf rust and leaf blotch each may cause relatively small reductions in yield; however, if both diseases occur together, the combined yield reduction is greater than the sum of the damage each would cause alone. Thus the vulnerability of a crop may be greatly increased if varieties grown are subject to damage by two or more diseases.

The increasing amount of air pollution is also a potential hazard to productivity of wheat in some areas. Generally speaking wheat is more resistant to air pollutants than some other crop plants; however, it is known that wheat varieties vary in their sensitivity to air pollutants and thus are genetically vulnerable to damage by them.

UNFORESEEN VULNERABILITY

Genetic vulnerability is sometimes unforeseen. Such instances in the past are rare, but, when they have occurred, the results have been catastrophic. A classic example is found in oats. The Victoria gene for crown rust reaction was bred into a number of oat varieties that were grown over extensive acreages. The presence of this gene permitted a fungus, *Helminthosporium victoriae* Meehan and Murphy, to parasitize and practically destroy a large part of the acreage of oats that carried the Victoria gene. This vulnerability was totally unexpected.

In wheat a linkage between adult-plant reaction to the stem rust fungus and a genetic defect known as "brown necrosis" has long been known to exist. The slight reduction in yield from brown necrosis was negligible as compared with the favorable effect of the control of stem rust in spring wheat varieties carrying these genes. However, when these varieties were grown in the high, cool, and moist valleys of Mexico, they were severely damaged by brown necrosis.

In most of the cases of catastrophic damage from certain "new" diseases, knowledge of the genetic vulnerability of the host did exist. The destruction of the American chestnut by chestnut blight resulted from the introduction into North America of the parasite, but the vulnerability of the chestnut to the parasite was known from research done in China. The vulnerability of corn with T cytoplasm to southern corn leaf blight was known prior to the epidemic of 1970 from work done in the Philippine Islands. The vulnerability of flax to wilt was well known in the first part of this century, and it was planted on new land until the soil became infested with the parasite and the farmer was forced to plant other crops or move on to virgin land. The flax crop was threatened as new land became scarce, but resistance was discovered.

Bruehl (Quisenberry and Reitz, 1967) after describing 23 diseases other than the rusts, smuts, and viruses of wheat, says "a myriad of spots and blights are omitted from this chapter. Probably nobody knows all the possible troubles of wheat." It is primarily this "myriad of spots and blights" that constitutes the unknown vulnerability of wheat to diseases.

Much but not all unforeseen vulnerability is eliminated in plant breeding programs. If vulnerability is carried it is usually because of linkage with a highly desirable character, for example: brown necrosis linked with adult plant reaction to stem rust of wheat, vulnerability to Victoria blight with reaction to crown rust of oats, and the vulnerability to leaf blotch with the dwarf character of the Mexican wheats. In these cases vulnerability did not show up until new varieties were moved to a new environment or a variant of a fungus developed and became widespread.

CONTROL PRACTICES

Control of diseases in wheat theoretically can be achieved by several approaches: wheat pathologists have long considered the possibilities of controlling diseases by genetic manipulation of the parasite. Some of the suggestions are "recycling" of host genes, developing multiline varieties, maintaining varietal diversity, restricting use of certain genes to limited geographical areas, and using certain other genes only in combination. Some of these suggestions have been tested, and preliminary results appear promising. The suggestion that we should accept a constant loss each year rather than a catastrophic loss occasionally by using general resistance and stabilizing selection perhaps is workable in some instances but at present may not be acceptable. The suggestion that cultures of a pathogen having many genes for virulence are less aggressive has been shown to be incorrect in some cases. Nevertheless, it is likely that certain genes or combinations of genes for virulence and/or reaction do indeed have a depressing effect on the development of disease.

Cultural practices that extend the growing season increase vulnerability to disease, and because of this wheats should be bred that reduce the period of vulnerability without reducing the beneficial effect of the cultural practices. An outstanding example is the development of winter wheat varieties for the central United States that mature 10 days earlier than old varieties and thus usually escape damage from the rusts.

More knowledge of the ecology of parasites, in particular the leaf blight and foot and root rot fungi, is needed. Application of such knowledge could lead to at least reducing, if not eliminating, damage from many wheat parasites.

Perhaps the best approach to control is an integrated one. We should continue to breed for specificity whether in mono- or multi-line varieties since, when it works, it is highly effective. However, specificity should be used only with wheats that are known to possess useful nonspecificity or general resistance because then a mutation to virulence in the parasite will not be as damaging. In addition, work should continue on the development of better fungicides to be used if all else fails. It is known that fungicides are more effective when nonspecificity occurs in the host-parasite association. It is evident, then, that nonspecificity is a key in an integrated program for control of damage from disease. We know so little about the phenomenon that some workers even question whether it exists. This lack of knowledge is perhaps the greatest barrier to overcoming the vulnerability of wheat to damage from diseases.

VULNERABILITY OF WHEAT TO INSECTS

Wheat has been plagued by insects since the beginning of recorded time. Before Christ, grasshoppers destroyed grain crops in Africa, Asia, and Europe. Outbreaks of Hessian fly (*Mayetiola destructor* [Say]), greenbug (*Schizaphis graminum* [Rond.]), wheat stem sawfly (*Cephus cinctus* Nort.), armyworms, cutworms, and grasshoppers have been recorded throughout the history of agriculture. Today, wheat is still subject to damage by more than 100 species of insects.

The cereal leaf beetle (*Oulema melanopus* [L.]), an enemy of small grains, continues to increase in numbers since it was first found in one county in Michigan in 1962. It is now found in 10 states and in parts of Canada.

Prior to 1970 the greenbug, an enemy of small grains in the United States, had never been destructive in the Pacific Northwest. That year a new strain of greenbug built up rapidly in wheat fields of south central Washington. Over 4,000 acres had to be reseeded and 40,000 acres were chemically sprayed before the outbreak was contained. Severe losses of wheat also occurred in the Republic of South Africa. In 1970 40,000 acres of wheat in the Eastern Orange Free State were treated with insecticides for greenbugs, and an aerial spray program prevented a total loss of wheat to this insect (Chemagro, 1971). Stored wheat is very vulnerable to damage by insects, rodents, moles, and fungi, particularly in areas of the world where storage facilities are inadequate or nonexistent. In certain areas of India where the development of proper storage facilities has not kept up with the increased grain production resulting from the new Mexican wheats, heavy losses occur in grain stored on the ground.

In the United States alone, the average annual loss to insects has been estimated to be $142 million (Quisenberry and Reitz, 1967). Also, indirect losses amounting to a million dollars annually are attributed to virus diseases transmitted by insects. The wheat plant is continuously being attacked by insects, but unless symptoms of an outbreak are visible, most of these losses go unnoticed.

CONDITIONS INFLUENCING VULNERABILITY

Whether or not a wheat crop becomes vulnerable to insect attack depends upon the genetic makeup of wheat plant, the genetic makeup of the insect, environmental conditions, and the interaction of all three. If the wheat crop provides an adequate supply of food and shelter and if environmental conditions are at an optimum for the insect, and if there is an absence of predators and parasites, then an outbreak is likely to occur.

Probably outbreaks are more dependent upon optimum weather conditions than any other factor. Many insect species require a dry, warm environment for survival, whereas others require a cool, moist environment. Spring wheats are not subject to as many insect pests as winter wheats because of their adaptability to a cold climate that is unfavorable to the development of many insect species. For instance, the extreme cold winters of North Dakota, South Dakota, and Montana do not permit the establishment of Hessian fly, whereas the wheat stem sawfly does thrive under these conditions.

Nutritional factors influence insect abundance. A new wheat variety may be more favorable nutritionally than an older variety, and an insect species may build up into epidemic proportions. On the other hand, a new variety could be poor nutritionally and the insects might not survive or their growth might be adversely affected with a resultant reduction in the number of insects.

Our agricultural practices today also influence insect abundance and distribution. The practice of planting large acreages of a crop that is favorable to a specific insect species hastens the establishment and abundance of this species. If, however, the wheat is resistant to the insect, its effect will be to reduce the insect population. Irrigating wheat land previously arid may change the environment to such a degree that additional insect species may move in, become established, and build up into epidemic proportions. A newcomer insect species, such as the cereal leaf beetle, will also have fewer of the natural enemies present that ordinarily would keep the insect population from increasing in size.

Heavy fertilization has the effect of producing more succulent growth of the plant that is favored by the insect for food and shelter, which might result in increased propagation of the insect species. On the other hand, more plant foliage and a more robust plant may better tolerate insect feeding and the plant may not have its yield affected at all.

Time of planting also influences the vulnerability of the wheat plant to insect attack. Prior to the development of varieties resistant to the Hessian fly, wheat was planted late on "safe seeding dates" so as to escape the emerging adults that normally laid their eggs on the fall-planted wheat at an earlier date. Planting late sacrificed yield increases, but it did prevent major outbreaks. However, as effective as it was in the fall, it did not prevent losses in the spring from Hessian fly populations that emerged from volunteer wheat that was infested the previous fall.

Volunteer wheat is also responsible for attack by the wheat curl mite, *Aceria tulipae* (Kiefer), resulting in severe outbreaks of wheat streak mosaic in Kansas and Nebraska. It has been shown that volunteer wheat resulting from hail damage early in the season is the source of mite populations that provide the source of inoculation for the seeded fall wheats (Gibson and Painter, 1956).

Although planting wheat late may be a device for protecting it from Hessian fly attack in the winter wheat region, early maturing wheats are responsible for the nonsurvival of the wheat stem sawfly in the spring wheat areas of northern Nebraska and southern South Dakota. Unpublished ecological studies conducted by USDA personnel in Montana and North Dakota have shown that the development of the sawfly and the wheat plant are not synchronized in these areas: the wheats mature before the sawfly completes development and hence the sawfly is starved out or the plant tissue becomes too hard for the sawfly to survive.

Along with the abundance of genetic diversity in the wheat plant for survival against insect attack, there also is a great amount of genetic diversity in the insect for surviving and infesting the wheat plant. The life cycle of the insect, its method of reproduction, its complex digestive and nervous system, its mobility, and its feeding behavior all add to its genetic diversity, making it a constant menace.

Insects feed on particular parts of the plant. There are root feeders (wireworm, false wireworm, white grub) leaf feeders (armyworm, cereal leaf beetle, grasshopper), stem and new shoot feeders (Hessian fly, wheat-stem sawfly). The wheat blossom midge feeds on the head and causes considerable damage.

With the exception of white grubs, wireworms, cutworms, and armyworms, most wheat insects attack only wheat or the closely related small grains, oats and barley. Many do not have other hosts, and if controlled on a particular crop will not be able to maintain themselves on another crop. However, such insects as the Hessian fly and wheat-stem sawfly can survive on certain native grasses, which makes the control of these pests more difficult.

GENETIC CONTROL OF INSECTS

For at least two insects, the Hessian fly and the wheat-stem sawfly, resistant wheat varieties have been the major method of control over the years and have proven very satisfactory in reducing populations of these insect species. Resistant varieties utilize the inherent genetic variability of the plant to its advantage; by the manipulation and transfer of genetic material from one plant to another, resistant varieties of good agronomic character can be developed.

It has been estimated that 20 million acres of wheat are in areas infested with Hessian fly, and of these 8.5 million acres were sown to Hessian fly-resistant varieties in 1969 (Gallun and Reitz, 1971). At a savings to the farmer of at least $2.00 an acre, $17 million has been returned to agriculture in one year by planting Hessian fly-resistant varieties. It has been estimated that the entire cost of developing all Hessian fly-resistant varieties to date amounted to approximately $6 million. One year's savings by planting resistant varieties more than offsets the entire cost of developing these varieties for the past 50 years!

Six states now grow wheat-stem sawfly-resistant wheats on approximately 1.5 million acres. Here again there is at least a $3 million return to agriculture in one year. Wheat varieties are now being developed in federal and state laboratories for the greenbug, the cereal leaf beetle, and the wheat-stem maggot. The USSR lists wheats that are resistant to the Hessian fly, and Australia and New Zealand are developing nematode-resistant wheats. Yugoslavia, under Public Law 480, is cooperating with the United States in developing cereal leaf beetle-resistant wheats.

Resistant wheats do have limitations, however. Wheats that are resistant to one insect species may be vulnerable to attack by another species. Also, a resistant wheat may not be equally growable in different areas of the country because of its limited adaptability.

Probably most significant is the mechanism of resistance utilized. Wheats are resistant to insects in three ways.

In *antibiosis,* the insect attacking the plant either dies or the plant affects the insect or its progeny adversely.

In *nonpreference,* the attacking insect moves from a given plant to a more attractive one for food, shelter, and egg laying.

Third, a plant having *tolerance* as a mechanism of resistance may be able to withstand a heavy insect attack and still produce a good crop.

Antibiosis is the principal method used, and always a few variants in the field have enough genetic diversity to survive on the wheats that ordinarily kill the rest of insects of this species. These variants, called races or biotypes, interbreed and eventually build up into epidemic proportions because of their ability to survive on the wheats that were heretofore resistant. This occurred with Hessian fly resistant wheats in Indiana (Gallun and Reitz, 1971).

Dual, the first Indiana wheat to show resistance to the Hessian fly, was released to wheat growers in 1955. It had a single resistance dominant gene, H_3. Between 1955 and 1962, Monon and Redcoat, also carrying H_3, were released. By 1962, these three varieties were growing on approximately 68 percent of the wheat acreage in Indiana. These wheats also had high yields and resistance to diseases.

A 1956 survey showed the value of Dual. Where actual or reasonable comparisons were possible, fields of Dual wheat had an average infestation of 1.2 percent, whereas susceptible wheats averaged 66 percent (Cartwright and Gallun, 1956). The H_3 gene was used because it was available, gave good protection, and was easy to transfer into good agronomic susceptible wheats. Other genes for resistance were also available and were being incorporated into newer wheats in case races of Hessian fly that could attack Dual, Monon, or Redcoat were to develop.

As was expected, a new race of Hessian fly did develop in a few fields, in 1962. An examination of infested plants and emerging adults showed the presence of a new field race, called Race B, that could attack wheats having the H_3 gene for resistance. These new

Hessian flies, because of their genetic variability, were able to build up into epidemic numbers in isolated areas.

A new variety of wheat, Knox 62, with resistance to Race B, was released in 1962. It had the H_6 gene for resistance to Race B. Benhur, another variety having the same kind of resistance as Knox 62, was released to wheat growers in 1966. To date, these two wheats have suffered very little infestation by Hessian fly, whereas other wheats having the H_3 gene have become heavily infested in a few isolated areas of Indiana. A new wheat, Arthur 71, having resistance to all known races of Hessian fly has been developed and was released to wheat growers in 1971; but the cycle may repeat.

A good example of nonpreference is the cereal leaf beetle, which does not lay eggs on hairy leaves even when the beetles are confined to the plant under cages (Schillinger and Gallun, 1968). There has been no natural selection for biotypes of beetles that can lay eggs on hairy leaves. Adults will move away from hairy leaves and lay eggs on wheats or grasses that are smoother; hence selection does not occur. In the event that large acreages are sown to hairy wheats, it will remain to be seen whether the insect changes its preference, or if from lack of food the cereal leaf beetle dwindles to numbers of noneconomic importance.

Beetle populations may survive on nonresistant oats and barley, however, if these crops are planted in the vicinity. Beetles not finding satisfactory wheat hosts may hold off laying eggs until an acceptable host such as spring-planted oats or barley is available. If this occurs, the solution is to cease planting oats or barley in the area of hairy wheats for a number of years, until the level of the beetle population drops. Developing oats and barley having the nonpreference mechanism is currently being investigated.

SUMMARY

From the foregoing discussion, it can be seen that vulnerability of the wheat crop to insect attack depends upon many factors all directly responding to the inherent variability of the wheat plant and its interaction with the insect, which also has a high degree of variability. Vulnerability to insect attack can be reduced by proper management of agricultural crops, the use of specific insecticides, and the utilization of resistant varieties.

Growing resistant varieties is an efficient and economical method of controlling insects, although this has its limitations. When developing resistant varieties, different genes for resistance should be utilized and mechanisms of resistance, nonpreference, or tolerance, if available, should be used in addition to antibiosis because of fewer chances to select for races or biotypes. Combining genes for resistance in a single plant may lessen the chance that plants will become vulnerable to new races. Using single gene resistance, and having other genes for resistance available in wheats for substitutes when biotypes develop, can also be used. New approaches to insect control should be pursued, and progressive research programs should be conducted to study the interaction of the host plant and the invading insect so that all is in favor of the plant and nothing favors the insect.

Although the wheat plant has a considerable amount of vulnerability to insect attack, the wheat plant has been in existence for 15,000 years and it is constantly being improved. Major widespread outbreaks of insect pests have not occurred where resistant varieties have been utilized and progressive breeding programs have been conducted. Vulnerability to insects will most likely occur with the introduction of a new pest from foreign countries or because of a continuous use of similar germ plasm in resistant varieties using antibiosis as the mechanism of resistance.

OUTLOOK

STRENGTHS

• Wheat is a widely adapted crop with a polyploid genetic makeup that permits chromosome engineering to a degree not possible in most other crop plants. Hence the crop represents a well-buffered genetic system. Whole or partial chromosome substitutions from related species have been accomplished to achieve resistance to diseases.

• Because it possesses several chromosome sets and has been grown in environmentally-diverse regions of the world for many centuries, wheat possesses a large amount of genetic variability. The many non-cultivated forms of wheat and related species have been intercrossed with cultivated wheat. Thus, there is a continuing flow of genetic material into the cultivated types. Wheat may be a highly receptive crop for creation of new variability by induced mutations because it can withstand genetic alterations as well as store and accumulate them. Genetic and chemical sterility systems are now available to accelerate genetic recombination in wheat and wheat relatives and thus increase genetic variability.

• Extensive germ plasm collections of wheat are in existence. The largest is that maintained by the USDA. A recent FAO germ plasm survey has determined that collections of wheat and wheat relatives are being held at 114 centers in the world. *Triticum* germ plasm alone totals 221,086 items. The viability and thereby the usefulness for breeding of some of the collections is not known, nor is the amount of duplication.

• Because of the worldwide importance of wheat as a food crop, there are many programs of wheat improvement, and the scientific manpower engaged in the study and solution of wheat problems is large.

WEAKNESSES

• Relatively few specific resistances to pathogens, insects, and other hazards are widely used in the United States and the world. Their continued use may render wheat more vulnerable to attack and disastrous losses.

• Virtual monoculture of wheat has developed in some wheat production areas of the world. Bezostaia, Gaines, Scout, and the CIMMYT semidwarf types, representing extreme potential genetic vulnerability, are examples.

• The importance of wheat as a food subjects breeders to strong pressures from the wheat-processing industries. These tend to have a restrictive influence on the germ plasm base of breeding programs.

• The erosion of centers of natural wheat diversity is occurring at an alarming rate. Comprehensive organized efforts to preserve such diversity are slow in development.

RESEARCH NEEDS

• The genetic variability of wheat is being rapidly lost. It is imperative that ways be found to offset the consequences of this trend. The diversity in centers of origin of wheat should be intensively sampled, collected, categorized, and stored. Additionally, wheat breeding programs, cytogenetic laboratories, and induced mutation laboratories should be considered centers of origin. Wheat collection efforts should be continuous and involve teams of qualified scientists from several disciplines working in field laboratories. Documentation and publication should be an important part of the total effort. Better systems for making information available must be developed.

• International living reservoirs of wheat germ plasm must be established and maintained, with sites including all of the major wheat production environments. Available genetic sterility systems should be incorporated in the germ plasm reservoirs and chemical male gametocides utilized to enhance genetic recombination.

• Wherever vulnerabilities of wheat occur in the world there should be developed without delay "on-site" continuous, competent research to deal with these problems.

• There should be encouragement and support of agricultural research organizations to investigate nonspecific characteristics of wheat that render the crop less subject to damage from biotic and environmental hazards.

• Nonpreference and tolerance mechanisms of insect resistance must be more extensively utilized in wheat in addition to antibiosis.

• At the present time diseases and insects represent the greatest vulnerability of existing varieties of wheat. Preoccupation with them may have led to lack of concern for vulnerabilities to environmental stresses and other hazards of wheat production. There is a very real danger that such preoccupation will cause breeders to be unconcerned with valuable inherent traits that are independent of pathogens and insects. Attention should be focused on dangers of this kind.

• There are many hypotheses concerned with lessening the vulnerabilities of wheat, and there is a tendency to assume that we have sufficient knowledge to cope with these vulnerabilities. There should be continuing effort to define and identify the vulnerabilities of wheat and devise genetic and management systems to minimize their danger to stable wheat production.

• Concentrated research on development of selective and biodegradable pesticides must continue. The research should include identification of systemic chemicals that could be utilized as wheat seed treatments with minimal danger to environment. The use of chemical pesticides and fertilizers is basic to productive wheat culture.

• The long-range effect of the Variety Protection Act on germ plasm development, preservation, and exchange should be carefully studied. Interpretations and procedures must be developed to assure continued free exchange of wheat germ plasm.

• Biological controls in addition to resistance mechanisms in wheat should be investigated and steps taken to implement their use when feasible.

REFERENCES

Borlaug, N. E. 1968. National production campaigns, p. 98 to 113. *In* The Rockefeller Foundation, Strategy for the conquest of hunger. New York.

Brock, R. D. 1970. The role of induced mutations in plant improvement. IAEA/FAO Symposium on Induced Mutations and Plant Improvement (Buenos Aires, Argentina), Proc.

Cartwright, W. B., and R. L. Gallun. 1956. Reaction of Dual wheat to Hessian fly attacks in 1956. U. S. Dep. Agric. Entomol. Res. Branch, Cereal and Forage Insects Sec. Spec. Rep. W–65.

Chemagro Corporation. 1971. Aircraft flew through clouds of aphids. Special report from Farbenfabriken Bayer AG. Chemagro courier (Chemagro Corp., Kansas City, Mo.). January.

Creech, J. L., and L. P. Reitz. 1971. Plant germ plasm now and for tomorrow. Adv. Agron. 23:1–49.

FAO. 1967. Report of the FAO/IBP Technical Conference on Exploration, Utilization and Conservation of Plant Genetic Resources (Rome, Italy).

Gallun, R. L., and L. P. Reitz, 1971. Wheat cultivars resistant to the Hessian fly. U.S. Dep. Agric. Prod. Res. Rep.

Gibson, W. W., and R. H. Painter. 1956. The occurrence of wheat curl mites *Aceria tulipae* (K) (Errophyidae), a vector of wheat streak mosaic in winter wheat seedlings grown from infested kernels. Kans. Acad. Sci., Trans. 59:492–494.

Johnson, T. 1961. Man-guided evolution in plant rusts. Science 133:357–361.

Loegering, W. Q. 1968. Problems of storing plant disease data for retrieval by automatic processing machines. Wash. Agric. Exp. Sta. Bull. 705.

Quisenberry, K. S., and L. P. Reitz [ed.]. 1967. Wheat and wheat improvement. Am. Soc. Agron. (Madison, Wis.) Agron. Monogr. 13.

Reitz, L. P., and J. C. Craddock. 1969. Diversity of germ plasm in small grain cereals. Econ. Bot. 23:315–323.

Rowell, J. B., W. Q. Loegering, and H. R. Powers, Jr. 1963. Genetic model for physiologic studies of mechanisms governing development of infection type in wheat stem rust. Phytopathology 53:932–937.

Schillinger, J. A., and R. L. Gallun. 1968. Leaf pubescence of wheat as a deterrent to the cereal leaf beetle, *Oulema melanopus.* Annu. Entomol. Soc. Am. 61:900–903.

Thurston, H. D. 1971. Relationship of general resistance: late blight of potato. Phytopathology 61:620–626.

Young, H. C., Jr. 1970. Variation in virulence and its relation to the use of specific resistance for the control of wheat leaf rust, p. 3 to 8. *In* Plant Disease Problems. First Int. Symp. Plant Pathol. Proc.

CHAPTER 10

Sorghum and Pearl Millet

Contents

SORGHUM 156
  Adaptation 156
  Utilization 157
  Hybrid Production 157
  Breeding 158
  Agronomic Practices 159
  Germ Plasm Resources 160
    Genetic Background; Sources of Resistance
  Parasites and Diseases 161
    Major Vulnerabilities; Introduction of Parasites;
    Future
  Insects 165
    Major Pests; Pest Interrelationships; Possible New
    Threats; Insect Adaptation
  Summary 167
PEARL MILLET 167
  Current Situation 169
  Germ Plasm Resources 169
  Diseases and Insects 170
  Avoiding Crisis Situations 171

155

SORGHUM

Sorghum is cultivated in the warm, subhumid and semiarid parts of the world. It is adapted on all continents in areas with 125 or more frost-free days, and where the summer temperature exceeds 20 C. Sorghum is a staple food item in many sections of the world; in other areas it is used primarily as feed grain or forage.

In 1969 sorghum occupied approximately 200 million acres worldwide. It ranks fourth in acreage of the crops of the world, exceeded only by wheat, corn, and rice, and the United States, Asia, and Africa are the major areas of production.

ADAPTATION

Sorghum is grown primarily in the Great Plains and Southwest in the United States because it is better adapted to semiarid conditions than other grain crops. While the crop performs better under more favorable moisture, temperature, and humidity conditions, it does have high resistance to desiccation; its extensive fibrous root system and a xerophytic leaf enable it to survive under adverse conditions. More

156

than one-fourth of the United States acreage is irrigated. Commercial fertilizers are used on two-thirds or more of the acreage.

UTILIZATION

Short sorghums that are adapted to the combine harvester are used for grain production in the United States. Taller and usually later-maturing types are used for forage and silage.

In 1969 13.5 million acres were harvested for grain, 2.8 million acres for forage, and 0.7 million acres utilized as silage. Average yield per acre for grain sorghum in 1969 was 55.2 bu. Total production was 743 million bushels, which represents a 50 percent increase over 1959. Forage yields have remained constant at around 2 tons per acre, silage yields have increased from an average of 8.6 tons per acre in 1959, to 11.7 tons per acre in 1969.

Texas, Kansas, and Nebraska account for 80 percent of the grain sorghum acreage in the United States. Texas leads in total acreage with 5.9 million harvested acres in 1970. Slightly over 500,000 acres are grown in Oklahoma, while other states with more than 200,000 acres are California, New Mexico, South Dakota, and Missouri. Kansas grows almost half the acreage used for silage; Texas is the major producer of sorghum for forage.

HYBRID PRODUCTION

Development and production of first generation hybrids in grain and forage sorghums was accomplished on a commercial scale in 1957, and seed was made available to farmers on a limited scale that same year. Producers rapidly converted to hybrids, and within a few years practically all of the grain sorghum acreage and most of the forage and silage areas were planted to hybrids. Since 1960 hybrids between grain sorghum and the forage types (and in more recent years crosses between sudan types) were offered to growers. These new crosses have enabled the acreage of this summer forage crop to expand significantly. Several major seed companies conduct extensive research programs with sorghums, and the bulk of the seed used in the United States is produced by private companies. Most of the hybrids currently being offered to growers appear to be of a parentage similar to that of earlier hybrids, but they stand better and yield more per acre. One major exception is the introduction and extensive use of varieties with yellow endosperm (the tissue that nourishes the developing embryo), that were introduced from Nigeria.

FIGURE 1 Harvesting sorghum. (Photo courtesy R. F. Holland, DeKalb Agricultural Research, Inc.)

Breeding work in the United States is focused on development of superior parents for hybrids. The major (or possibly the only) source of cytoplasmic male sterility used by seed producers is the milo source discovered in 1954 by the Texas and Nebraska Agricultural Experiment Stations. Other sources of male sterility have been studied in a preliminary way. The original source has been stable over a wide range of conditions.

BREEDING

Strong emphasis in all sorghum research programs is placed on incorporation of disease resistance. Sources of resistance have been determined for several major diseases, and some of these can be incorporated through simple backcross procedures. Inheritance of resistance to such important diseases as charcoal rot is complex, and additional breeding methods must be used to develop tolerance. Considerable attention has been devoted to determining sources of resistance to various insects, but incorporation of high degrees of tolerance has not been accomplished readily. However, most of the insects can be controlled with proper applications of insecticides, so lack of insect resistance has not presented an insuperable problem to growers. Discovery in recent years of a source of greenbug resistance offers considerable promise since this is one of the major pests of sorghums in the Great Plains and Southwest.

Discovery of high-lysine corn has heightened interest in developing sorghums with better nutritional qualities. A parallel high-lysine type to that discovered in corn has not yet been isolated in sorghum, but some data indicate that higher levels might be obtained. Yellow endosperm types are also available that offer some benefit by increasing xanthophyll and carotene pigments in the grain. Carotene in yellow grain sorghum is typically one-half or less than that normally found in yellow corn. Other grain types such as waxy seed varieties have been used to a limited extent for special industrial usage. Less than 1 percent of the acreage is used for syrup production.

AGRONOMIC PRACTICES

Yields per acre experienced by farmers in the United States since the advent of hybrids have increased twofold. Twenty to 30 percent of the increase may be attributed to hybrid vigor, and the balance to increased number of acres grown under irrigation, use of more fertilizer, and other improvements in agronomic practices. Irrigated grain sorghum typically yields three to four times more grain than dryland crops grown in the same area. Surface irrigation is the most common method used although sophisticated overhead systems are becoming more evident. The use of fertilizer on sorghums has been increasing rapidly in recent years and heavy rates of nitrogen are commonly used on irrigated fields. Anhydrous ammonia is the most common form of fertilizer used, though balanced formulations are applied in many areas. Sorghums respond readily to fertilizer. Twenty-inch row spacing is known to give yield advantage over 40-in. spacing and this procedure, or some such variation as planting 2 rows 12 in. apart on 40-in. centers, is gaining wider acceptance. Herbicides are available for pre- and postemergence application, and their use is growing. Plant populations per acre are increasing, particularly under irrigation. Sorghums will tiller and compensate for poor stand density to some extent, but most growers prefer to plant heavier rates in order to obtain a higher degree of uniformity at harvest time.

GERM PLASM RESOURCES

Sorghum has been in the United States since 1800 having apparently reached the Western Hemisphere from the Guinea Coast of Africa during the time of the slave trade. By 1905 farmers had changed tall, tropical varieties into short-statured, temperate varieties by finding and preserving mutations. Before World War II plant breeders had achieved varieties with still shorter stature and made sorghum suitable for mechanical harvesting.

Genetic Background When sorghum research workers found sterility-inducing cytoplasm in milo, they had then no obvious reason to look for other sources. Sterility-inducing cytoplasm was found in several grass sorghums, but their cytoplasms must be somewhat similar to milo cytoplasm because they were identified by using the same genes that workers had used to discover the original cytoplasmic male sterile. Six of these cytoplasms have recently been released to sorghum breeders. No other effort has been made to find more cytoplasmic-genetic systems of male sterility and restoration, though such systems can probably be found by using the same method used to find the present system.

The Texas Agricultural Experiment Station and the USDA have recently released 62 short-statured, temperate-zone varieties that were originally tropical varieties. Their cytoplasms have not been characterized and it is possible that other sterile cytoplasms and sterile genes may be found among them.

Plant breeders who developed short-statured varieties suitable for mechanical harvesting with a combine in the 1920's and 1930's used milo and kafir as parents. Older varieties of taller stalk height carried, between them, the necessary genes to produce the short-statured characteristic. For that reason, practically all of the parents of the first hybrids put into production were selections from milo and some kafir. As breeding work continued to find parents of new hybrids, the original milo and kafir derivations have been crossed to feterita, hegari, or some derivative of the yellow endosperm variety, Kaura. A few new females are in use, but kafir is one of the parents of all of these. So, in addition to having only one kind of sterility-inducing cytoplasm, present parents of grain sorghum hybrids have at most five to six varieties in their parentage; and most parents have only two or three. Thus, the genetic diversity in parents is not sufficient to give adequate protection against a catastrophic epidemic and a real threat exists.

Sources of Resistance Because sorghum found its place in the sub-humid or semiarid area of the United States, it is generally not recognized that sorghum grows in humid climates in the tropics. However, sorghum varieties exist that are resistant to most diseases and insects of humid regions as well as those that are problems in drier climates. Many sorghum varieties of the world have been brought together into a world collection, and resistance to certain insects and diseases that attack sorghum is available.

Numerous potential female parents exist among the varieties in the world collection. Any that appear promising, especially those that are resistant to several of the important diseases, should be converted to short stature and temperature-zone adaptation. In this way, the danger inherent in too narrow a genetic base might be reduced.

In view of the apparent uniform resistance to diseases that exist in many sorghum varieties in the world collection, it appears that the sorghum crop is as well protected as could be hoped for. But more of the resistant varieties need to be identified and converted to temperate zone adaptation. The identification of the desirable varieties, rather than the conversion to temperate adaptation, is the vexing aspect of the problem.

The identification of disease- and insect-resistant varieties in the world collection is important to the future of grain sorghum production in the United States. But the work of identification cannot be done in the United States because most of the resistant varieties are tropical and because many of the threatening diseases and insects are, fortunately, not in the Western Hemisphere. The need to identify desirable varieties justifies considerable effort and expense, and the work could probably be done best in tropical Africa.

The greatest threat to the crop lies in the milo cytoplasm in all female parents and hybrids at present, and additional sterile cytoplasms need to be identified. Parents with diverse genetic backgrounds must be developed, and one easy way to obtain new germ plasm is to adapt desirable tropical varieties to temperate-zone conditions.

PARASITES AND DISEASES

Sorghum became the nation's second most important feed grain not only because of its adaptability to warmer, drier climates but also because of its relative freedom in such climates from attack by pests and diseases. It is a fundamental principle of biology that, as cultivation of a given plant species becomes intensified, its vulnerability to pests and diseases increases. Sorghum is no exception. Most original introductions of sorghum to the United States occurred between 1853 and 1906, a period that was marked by few disease problems. By 1920, however, several varieties had been brought to a high state of purity, uniformity and productivity; sorghum acreages increased greatly, and in 1924 periconial root rot (caused by *Periconia circinata*) appeared as a new disease that became epidemic to milo sorghums. Concurrently, covered kernel smut (caused by *Sphacelotheca sorghi*) became an epidemic disease. The 1930's brought prolonged periods of hot, dry weather, and then the weak-neck disease and stalk rots aroused concern. Under intensified cropping practices during World War II, seed rot, damping-off, and seedling blight gained prominence. Most of those diseases were minimized through resistance, seed treatment, or cultural practices.

In 1954 the second century of sorghum culture in the United States began with the application of cytoplasmic male sterility to produce sorghum hybrids. Those hybrids increased production, but they also brought susceptibility to old diseases and the appearance of new diseases. Head smut became serious, especially in southern Texas, soon after large acreages were planted to newly developed hybrids. A single source of specific resistance was utilized, selection pressure was placed on the parasite, and a new race appeared in 1966. Maize dwarf-mosaic also appeared in 1966, again in Texas, and became epidemic on the Plains in 1967. The threat of mosaic was intensified in 1968 with the appearance of a new greenbug biotype (*Schizaphis graminum*), which was more efficient as a vector than corn-leaf aphid (*Rhopalosiphum maidis*), previously the principal vector. Tropical downy mildew (caused by *Sclerospora sorghi*) appeared in 1961 and became severe in southern Texas by 1965. None of these problems, however, has been identified with susceptibilities imparted by cytoplasmic male sterility *per se.*

Major Vulnerabilities The potential for epidemics in sorghum similar to the 1970 southern corn leaf blight exists. This would be made possible by parasites able to increase logarithmically in a relatively short period of time and to capitalize on specific germ plasms. Most fungal parasites of sorghum having such characteristics require more or less sustained periods of daily rainfall, dew, or high relative humidity.

Some do not. Moisture requirements of the former are less apt to be met in most grain sorghum regions than they are in the Corn Belt.

The disease-producing potential of *Sclerospora sorghi* is high. The fungus is introduced into new areas by sexually produced resting spores (oospores), and even if only a few seedlings become systemically infected from these, secondary production and spread of asexual spores (conidia) could become extensive with sustained humid weather.

Foliage diseases caused by bacteria or by fungi have been serious only in humid climates, but conceivably some representatives (*Helminthosporium, Colletotrichum, Gloeocercospora, Cercospora, Ramulispora,* and *Ascochyta* spp.) might produce diseases that could thrive sporadically on sorghums in the Great Plains.

Two fungus diseases that do not require excessively humid weather are head smut (caused by *Sphacelotheca reiliana*) and rust (caused by *Puccinia purpurea*). *S. reiliana* has limited potential to cause epidemics because only one generation is produced each season, but serious losses from head smut probably will occur periodically as long as specific resistance is used for control. Rust is usually associated with cool weather and maturing plants. Development of a rust fungus strain capable of reproducing under warmer temperatures may be the only event needed to make rust a prominent disease threat.

Humid climate is also not required for development of maize dwarf-mosaic. Johnson grass serves as a reservoir of primary inoculum in most sorghum regions. The virus of maize dwarf-mosaic can then be spread rapidly by aphids; it infects sorghum at most stages of maturity.

Most soil-borne parasites that cause root and stalk rots are non-specific, possibly because sorghum has evolved natural resistance to most root-stalk organisms. That resistance is effective under all but the most adverse growing conditions. It is possible, however, that such generalized resistance could be lost with continued narrowing of the genetic base for the crop. Some indication of this has been noted for *Fusarium moniliforme.* Specificity for host genotypes has been clearly demonstrated in red rot (caused by *Colletotrichum graminicolum*) and Milo disease (caused by *P. circinata*). Thus far, serious red rot losses have been restricted to warm, humid climates. Milo disease has been controlled with specific resistance, but a new strain of *P. circinata* that attacks most hybrids was identified in 1970 in the lower Colorado River Valley near Yuma, Arizona.

Introduction of Parasites Introductions of parasites from other continents, as occurred in the case of tropical downy mildew, is difficult to prevent. Ergot (caused by *Sphacelia sorghi*) has the potential for devastating effects in hybrid seed production, but some differences in resistance to ergot among male steriles have been reported in India. Other downy mildews (caused by *S. sacchari* and *S. philippinensis*), and possibly long smut (caused by *Tolyposporium ehrenbergii*) may have some potential for epidemics of major importance should they be introduced to the United States.

Future There is danger in being complacent about the present status of sorghum hybrids and their parental lines. For the near future of major food crops at least, preventive action against the possibility of catastrophic disease or insect problems before they occur should be considered seriously. Historically, such problems have been solved after the fact, whether they caused widespread famine and death, or whether they simply raised economic barriers to profitable crop production.

Practices in plant breeding seemingly have changed little because of epidemics that occurred as a direct consequence of genetic manipulations. Epidemics caused by new parasites or by new virulent strains of old parasites have always come as a shock, and even embarrassment, to our pride in husbandry or technology. Traditionally we have based foresight more on successes than on failures of past efforts. Potential threats of known pests and parasites should not, therefore, be evaluated solely on basis of their inherent capacity to do damage to cultivars that may have been made genetically vulnerable through breeding methods. Perhaps the future strategy of crop improvement might be changed so as to allow more concern for long-range effects of genetic manipulations.

Present genetic manipulations with simply inherited characters should be backed up with efforts to discover, retain or recapture naturally evolved, multigenic resistances to pests and pathogens. Maintenance and improvement of world collections are indispensable. This may require renewed plant explorations and establishment of all or part of the world collection at several additional key locations. Careful evaluation at sites most conducive for the development of each potential pest or disease is in order for most if not all such materials. Particular emphasis at this time should be placed on an evaluation of those known to possess natural resistances on exotic converted materials, sources of cytoplasmic male sterility, and on other germ plasms common to a majority of hybrids now in production. Artificial screening techniques should be developed or implemented as warranted.

INSECTS

Insects cause an estimated 7.5 percent loss in potential production of sorghum in the United States, Mexico, and Central America. This compares with 10 percent losses attributed to insects in both Asia and Africa, which combined have about 75 percent of the world's sorghum acreage.

Major Pests About 100 insect species throughout the world are pests of sorghum; less than a dozen species account for nearly all of the losses in the United States. In North and Central America the following species cause the major losses attributed to insect pests with the percent loss ratios indicated: corn earworm, *Heliothis zea* (45.6 percent); sorghum webworm, *Celama sorghiella* (15.6 percent); chinch bug, *Blissus leucopterus* (14.4 percent); sorghum midge, *Contarinia sorghicola* (12.2 percent); cutworms (5.6 percent); corn leaf aphid, *Rhopalosiphum maidis* (4.4 percent); southwestern corn borer, *Zeadiatraea grandiosella* (2.2 percent). Some other insects reported as pests of sorghum in the United States are fall armyworm, *Spodoptera frugiperda;* sugarcane borer, *Diatraea saccharalis;* sugarcane rootstock weevil, *Anacentrinus deplanatus;* and Banks grass mite, *Oligonychus pratensis.* Since 1968 the greenbug (*Schizaphis graminum*) has been the most important insect pest of sorghum. In that year nine states reported over 7.5 million acres infested with losses estimated to exceed $68 million.

Pest Interrelationships The host ranges of insects mentioned above (except corn earworm, fall armyworm, and cutworms) generally are limited to the Gramineae. From the economic standpoint, only the sorghum midge and sorghum webworm are primarily limited to sorghum. Since most insect pests of sorghum also attack corn, millets, and small grains, studies on the interrelationships among the various hosts are needed. For example, the greenbug is a threat to both wheat and sorghum, and in the Great Plains the growing seasons for these overlapping crops may provide favorable host plants throughout the year. An effective control on one crop may help protect the other.

Possible New Threats   Some examples of insects that are important pests of sorghum in Europe, Asia, or Africa, but not in the United States are: sorghum shoot fly, *Antherigona varia;* sorghum shoot bug, *Peregrinus maidis;* spider mite, *Oligonychus indicus;* mealy bug, *Heterococcus nigeriensis;* and several species of stem borers belonging to the Pyralidae and Noctuidae. The introduction of these or other pests into the United States could threaten our sorghum production. Where genetic resistance is available in other countries, these lines should be converted to types adapted in the United States. This would broaden the genetic base and make resistance more readily available if and when these pests are introduced. The parasites and predators of these pests should also be studied to see whether they might be rapidly introduced if needed.

Insect Adaptation   Sorghum has a wider range of adaptation than do some of the pests that attack it. For example, sorghum webworm is mainly a problem in areas having over 30 in. of rainfall. On the other hand, Banks grass mite causes damage primarily in semiarid regions. Natural selection and adaptive mutation may eventually enable these pests to extend their range. Research to develop control measures for one region may ultimately apply to others and consequently should receive broad support. When resistance to a pest that is limited to one region is discovered, it should be transferred to lines adapted to adjacent regions. Even if not used immediately, these resistant lines could be held in readiness and used as the need arises.

Studies on intraspecies variation of insect pests, especially the aphids, are indicated by the recent outbreak of greenbug on sorghum. The greenbug attacking sorghum, designated as biotype C, is different in several respects from types previously known as pests of small grains. Greenbugs from over the world vary considerably, particularly in host range, so the C biotype may have existed elsewhere prior to the United States outbreak. Other aphids, such as the yellow sugarcane aphid (*Sipha flava*) that attack the Gramineae should be evaluated for their potential as major pests of sorghum.

The greenbug problem on sorghum was not predictable, and it seems unlikely that we will anticipate the new sorghum pests of the future. Probably the best insurance against sustained losses from new insect pests is to make the genetic and cytoplasmic base as broad as feasible and to collect and maintain in seed storage all available germ plasm for future needs.

SUMMARY

The grain sorghums grown in the United States originated mainly from the kafirs and milos and to a limited extent from the hegaris, feteritas, and shallus. The kafirs and milos have been good parents for the development of productive varieties and hybrids, but are known to be susceptible to a number of diseases and insects found in this country and elsewhere in the world. Since grain sorghums have become a major crop in the United States, research scientists have been constantly involved in problems associated with the increase in prevalence of insects and diseases. Sorghum breeders have long recognized the genetic vulnerability of this crop in the United States because of the narrow germ plasm base being used. More recently the association of southern corn blight with T cytoplasm in hybrid corn has pointed out the devastating effect that a similar association of a disease and cytoplasm would have upon the grain sorghum industry.

Sorghum workers began exploring the uses of other sources of germ plasm a few years ago, and good progress has been made in transferring the germ plasm from tropical sorghum varieties into short-statured, day-neutral plants that can be used in the breeding programs in the temperate zones. Other breeding schemes have been formulated that will incorporate a random sample of the germ plasm found in the World Sorghum Collection into breeding populations. A start has been made to classify the few thousand items in the world collection not only for their morphological characteristics but also as to their reaction to the numerous insects and diseases found in the world. These projects, together with studies to be initiated to find other types of male-sterile cytoplasm and sterility genes (or other practical mechanisms for the production of hybrid sorghum seed), will aid in assuring the stability of this crop in the United States and elsewhere.

PEARL MILLET

Pearl millet, *Pennisetum typhoides,* a robust annual bunchgrass, occupies more than 45 million acres of the earth's surface and occurs in every continent. Although best adapted to the tropics, pearl millet also does well in hot areas of the temperate zone. It will grow and mature seed on sandy or rocky soils too acid, too dry, and too infertile for sorghum or corn. Yet it has great yield potential and, given a favorable environment, will equal or surpass corn and sorghum in forage production. Although the grain-production potential of pearl millet has not been well established, there is some evidence to suggest that it may also compare favorably with sorghum and corn as a grain crop.

Pearl millet has many uses. In the southeastern United States it serves as a forage crop. At all stages of growth it is free of the potentially hazardous cyanogenetic glucosides sometimes found in sorghums and when properly managed, millet generally surpasses other warm-season grasses in quality. In one experiment, steers that grazed on Gahi-1 pearl millet in Georgia gained over 2 lbs per day and produced over 500 lb of live weight gain per acre per year. High yields of top-quality silage have been obtained from boot-state Gahi-1 millet supplemented, when ensiled, with citrus pulp or ground snap corn.

Pearl millet is used primarily for human food in Africa and India. In areas where it serves as the principal grain crop it is usually preferred to other cereals and commands a premium at the market place. There many people consider it an unusually good food for the winter months, and some of them also believe that it is superior to other cereals as food for pregnant women. Chemical analysis of a number of Indian foods shows dehusked pearl millet seeds to be higher than rice, wheat, or corn in fat and minerals (particularly calcium and iron) and comparable in other principal constituents. This analysis also shows the content and balance of essential amino acids in pearl millet to be equal or superior to those of other adapted cereals.

More recent investigations indicate that pearl millet contains more protein and oil than wheat, corn, sorghum, or rice. Except for lysine deficiency, pearl millet has an excellent amino acid profile. Rats fed unsupplemented diets of pearl millet, sorghum, and corn grains made the highest growth rate on pearl millet.

The stalks of pearl millet left after the grain is harvested are used in Africa to build windbreaks, shelters, screen walls, and fences. In Africa and India, where fuel is short, people cook food over tiny fires fed with pearl millet stalks. Fodder left in the fields in these countries, although of poor quality, is usually consumed by cattle and goats during the winter months.

CURRENT SITUATION

In the United States Starr millet and an earlier maturing common type are the only open-pollinated varieties in use. More than half the acreage planted for forage is occupied by Gahi-1 (a first generation chance hybrid involving four inbred lines) and commercial single crosses, all of which probably use Tift 23A or Tift 23D cytoplasmic male steriles as the female parent..

India now has a sizable acreage of single-cross hybrid pearl millet, most if not all of which uses Tift 23A or Tift 23DA as the female. The percentage of India's 27 million acres planted to hybrids is not known. Some 20 percent of the increased food production in India associated with the "Green Revolution" has been credited to pearl millet hybrids.

Pakistan and Africa, the other major consumers of pearl millet, still use open-pollinated varieties that vary greatly in type. The A_1 cytoplasm of Tift 23A is the only one in general use. Another cytoplasm, A_2, carried in $239A_2$ has been released. A_2 produces excellent male steriles, and researchers are currently sterilizing several outstanding millet lines that are good maintainers for A_2 cytoplasm.

GERM PLASM RESOURCES

There is good reason to believe that pearl millet originated in Central Africa. Nigeria has two major types: the Gero (day-neutral) varieties, and the Maiwa (short-day) varieties, both occupying approximately equal acreages in Nigeria. These two types under different names are also grown in most of the African tropics. The short-day types have not been used in the rest of the world, probably because they will not mature seed before frost. The variety Tiflate represents the first attempt to utilize short-day types commercially outside of Africa. Some evidence suggests that the short-day types are more pest resistant than the day-neutral varieties.

Pearl millet possesses genetic variability similar in appearance to corn and sorghum. Much less of its germ plasm has been brought together in one place, however. We have about 500 introductions. It is extremely difficult to maintain introductions of this highly cross-pollinated species, and some collections have been lost as a result of sterility or outcrossing.

Pearl millet has generally produced more forage than sorghums on the sandy soils of the Southeast. It is more disease and pest resistant, is free of cyanogenetic glucosides, and appears to be comparable in forage quality. The promotion of commercial seed companies selling hybrid seed in this area will most certainly result in an increased use of this crop for forage.

The potential of pearl millet as a grain crop in the United States has not been ascertained. USDA workers at Tifton, Georgia, have developed one dwarf-grain hybrid that is to be compared with sorghum hybrids in several locations in the United States in 1971. Since pearl millet replaces other cereals in the drier, sandier, hotter areas of Africa and Asia, one might expect it to be most useful in similar habitats in the United States.

The excellent grain quality of pearl millet, and the preference that wild birds show for it, suggest that it can become an important grain feed for such animals as poultry if yields are satisfactory.

DISEASES AND INSECTS

Pearl millet appears to suffer from fewer pests, diseases, and insects than sorghum. In the United States *Pyricularia grisea* is perhaps the most serious foliage parasite. In Africa and Asia greenear disease, caused by *Sclerospora graminicola,* is a serious foliage disease and may kill susceptible material in very young growth stages. Head smut, caused by *Tolyposporium penicillariae* occurs around the world in the more humid areas where millet is grown. Ergot, caused by *Claviceps,* is very poisonous and is reported to cause death in animals consuming grain infected with it. It occurs in Africa and India, but has not been reported in the United States. Generally, diseases are not prevalent in the arid regions where millet is more frequently grown. Lines with a high degree of resistance to *Pyricularia grisea* and *Sclerospora graminicola* are available.

In the United States the fall armyworm, the sorghum webworm, the European corn borer, the corn earworm, and the lesser corn stalk borer attack pearl millet. These insects occur sporadically and rarely cause great damage. The midge and shoot fly that attack pearl millet in other countries have not been observed here.

The great morphological diversity exhibited by pearl millet suggests that resistance to its pests could be found if the world's germ plasm could be collected and screened. Developing good fertility restoration and reducing the time lapse between stigma and anther exsertion in this grass would reduce the incidence of smut and ergot.

AVOIDING CRISIS SITUATIONS

• Through a backcrossing program, transfer elite pearl millet lines into other sterile cytoplasms. We currently have two excellent sterile cytoplasms, A_1 and A_2. Other sources should be sought and used. Tift 23B is an excellent inbred line. It is quite likely that it will maintain sterility in more than one sterile cytoplasm, a possibility that should be explored. If Tift 23B will maintain sterility in cytoplasms A_1, A_3, and A_6, for example, Tift $23A_1$, Tift $23A_3$, and Tift $23A_6$ should be used in about equal amounts but in separate seed-production fields to produce Tift 23 hybrid seed. This appears to be better than blending all three, as some have suggested.

• Collect an array of maintainer and restorer lines to test sterile plants in the world collection for sterile cytoplasm type.

• Build up materially our meager world collection of germ plasm, screen it for pest resistance, catalog it, and put some of it in storage. Transfer genes for pest resistance to elite lines and populations.

• Convert the short-day types of pearl millet to day-neutral types, and vice versa, so that the desirable traits of each can be used interchangeably in breeding programs.

• Conduct basic research to develop alternative systems of producing hybrid seed without utilizing one or a few sources of cytoplasmic male sterility.

• The work suggested above, particularly the third and fourth items, could best be carried out at an international pearl millet center, and it seems obvious that such a center (adequately financed) should be established in the near future.

CHAPTER 11

Rice

Contents

INTRODUCTION	173
VARIETIES	174
COLLECTION AND PRESERVATION OF VARIABILITY	175
RICE IN THE UNITED STATES	175
Pests and Diseases	175
Weeds; Fungi and Nematodes; Insects	
Quality Characters	177
Genetic Uniformity	177
Variety-Related Diseases	179
DISEASES AND PESTS IN THE TROPICS	181
Crop Protection	183
Sources of Disease and Insect Resistance	184
Breeding for Resistance	185
Genetic Vulnerability of High-Yielding Varieties	186
Is a Common Gene for Dwarfism Dangerous?	188
SUMMARY AND CONCLUSIONS	188

INTRODUCTION

The genus *Oryza* is highly variable and has a worldwide distribution. It feeds half the world's people. There are about 20 well-defined species, two of which are cultivated. *O. glaberrima,* grown in a few African countries, is gradually being replaced by *O. sativa.* No serious effort is being made to improve *O. glaberrima,* and it probably will go out of cultivation in the future. The vast number of rice varieties exhibit an even vaster range in economically important characters.

From its tropical Asian origin rice has spread from latitude 40° south to 44° north and is grown from sea level to over 2500 m elevation.

Rice is typically irrigated, but it is also cultivated in rain-fed areas; where sometimes the water may exceed 3 m in depth. Rice is also grown as an upland crop without standing water, and is both transplanted and seeded directly.

These diverse conditions of geography, climate, and culture, coupled with the force of natural selection operating on the great reservoir of variability in *O. sativa,* resulted in much of the present varietal diversity. Man has also contributed to variability by develop-

173

ing cultural systems and selecting those varieties best adapted to each area.

VARIETIES

The varieties of rice are broadly classified into three groups: Javanicas, Indicas, and Japonicas. Javanicas are confined mainly to portions of Indonesia and have not been actively improved or used in improvement of the other two major groups.

Indica rice, found throughout the tropics of the world, is represented by thousands of varieties that have developed under an alternating monsoon-dry season climate, in areas of poor drainage, low solar radiation, high temperatures, infertile soils, and intense weed competition. This resulted in a basic tropical plant type characterized by rapid and vigorous seedling development and initial nutrient uptake, long and drooping leaves, tall and weak culms, profuse tillering, day-length sensitivity, late maturity, and grain dormancy. This varietal type has only a limited response to applied fertilizer or improvement in other cultural practices. The evolution of a relatively uniform plant type in tropical Indicas did not itself result in a narrowing of germ plasm. Thousands of varieties, each narrowly adapted, persisted; none has been widely grown.

The Japonica group, limited to temperate zones and the subtropics in Taiwan comprises hundreds of varieties. They evolved under lower, more evenly distributed rainfall, controlled irrigation and drainage, higher solar radiation, moderate temperatures, fertile soils, and relatively less weed competition. These forces, combined with active human selection, resulted in a distinct plant type having less seedling vigor, slower initial nutrient uptake, small and erect leaves, short and sturdy culms, moderate tillering, early maturity, and little or no grain dormancy. Japonica varieties have a marked response to cultural practices and possess a high yield potential.

Since the three groups of rice have been separated geographically, breeders find a partial sterility barrier when varieties from different groups are brought together and hybridized. Within each group similar barriers developed to a lesser extent. This has aided the preservation of variability. Some crosses between Indicas and Japonicas have yielded commercial varieties, particularly in the United States.

COLLECTION AND PRESERVATION OF VARIABILITY

The germ plasm of the 20-odd species of *Oryza* has been fairly well collected and is maintained at the International Rice Research Institute (IRRI), Philippines; the National Institute of Genetics, Misima, Japan; and the Central Rice Research Institute, Cuttack, India.

Over 14,000 accessions of cultivated rice are maintained at IRRI. The majority of these have been described and a varietal catalog is available. Seed of all accessions is held under long-term storage and is available to all rice workers. The collections in India, Japan, and the United States are somewhat smaller.

New collections are being made in the remote areas of Assam by Indian workers. Representative collections from Burma, Cambodia, Laos, and Vietnam are lacking, but plans are made to collect in these countries. Germ plasm from China should be collected and made available.

RICE IN THE UNITED STATES

Rice is a major crop in some parts of Arkansas, Louisiana, Mississippi, Texas, and California. In 1970 750,600 hectares (1,814,000 acres) were harvested in the United States. The farm value of the rice produced in the United States that year was $421,367,000. In 1969 rice ranked eleventh in farm value among all cereal, hay, vegetable, fruit, oilseed, and miscellaneous crops.

All rice in the United States is flood irrigated, so it is grown only where the land is reasonably level and irrigation water is available. Moreover, it is grown only in areas where the frost-free period exceeds 200 days and where the average daily temperature during most of the growing season is about 27 C. In the southern area rain commonly occurs throughout the growing season and the relative humidity of the air is fairly high. The moisture and temperature conditions are usually favorable for many rice diseases. The long planting season in Louisiana and Texas makes it possible to dilute exposure to insects and diseases. Since little rain falls during most of the growing season in California few diseases of rice occur there.

PESTS AND DISEASES

Changes in varieties and cultural practices have brought about subsequent changes in the prevalence of diseases, insects, and weeds.

For example some of the present varieties are more resistant to common physiologic races of some disease organisms than were older varieties. Short-season varieties now grown probably do not recover from rice water weevil (*Lizzorhoptrus oryzophilu*) attack as well as formerly grown long-season varieties. On the other hand, short-season varieties may partially escape injury from stinkbugs.

Weeds Weed-control practices now used have reduced the losses caused by weeds. They do not, however, control all species of weeds with equal efficacy. We know relatively little about the comparative ability of different genetic lines of rice to compete with weeds, or to tolerate different weed control treatments. Weeds that are particularly troublesome in rice include: barnyard grass (*Echinochloa crus-galli*), ducksalad (*Heteranthera limosa*), hemp sesbania (*Sesbania exaltata*), morning glory (*Ipomoea* spp.), northern joint vetch (*Aeschynomene virginica*), goose grass (*Eleusine indica*), red rice (*Oryza sativa*), sedges (*Cyperus* and *Fimbristylis* spp.), baronet grass (*Echinochloa* spp.), algae (blue-green and filamentous green spp.), bulrushes (*Scirpus* spp.), cattail (*Typha* spp.), and sprangletop (*Leptochloa* spp.).

Fungi and Nematodes Principal rice diseases caused by fungi in the southern states are blast (caused by *Pyricularia oryzae*), brown leaf spot (caused by *Helminthosporium oryzae*), narrow brown leaf spot (caused by *Cercospora oryzae*), seedling blight (caused by a number of fungi), seed rot (caused by *Achlya klebsiana* or *Pythium* spp.), bordered sheath spot (caused by *Rhizoctonia oryzae*), stem rot (caused by *Leptospheria salvinii*), and kernel smut (caused by *Neovossia barclayana,* sometimes called *Tilletia barclayana*). Other fungal diseases of rice in the United States include kernel spots that are caused by a number of fungi, leaf smut (caused by *Entyloma oryzae*), and stack burn (caused by *Trichoconis padwickii.*) A failure to set seed that may be caused by fungi is frequently observed in the southern states. Straighthead, a physiological disease, occurs on some soil types, and white tip, caused by a foliar nematode (*Aphelenchoides besseyi*), was formerly responsible for an important disease of rice in the United States. Early seeding, water seeding, and use of resistant varieties are factors in the disappearance of white tip. Hoja blanca, a virus disease that is transmitted by the rice delphacid (*Sogatodes orizicola*), has been reported in the southern states. In California seed rot and other seedling diseases and stem rot are the only diseases reported.

PLATE 4   Inspecting foundation seed of Northrose rice in Stuttgart, Ark.
(Photo courtesy Plant Science Res. Div., ARS, USDA).

Soil nematodes are reported to attack rice in Louisiana and Texas, although the loss caused by this pest is not clearly established. The rice root nematode, *Hirschmania oryzae,* was the only one of those present in the Gulf Coast previously known to damage rice, but recent investigations show that the ring nematode, *Criconemoides onoensis,* is also injurious and widespread.

Insects The principal insect pests of rice in the southern states are the rice water weevil and the rice stink bug (*Oebalus pugnax*). Other insect pests of rice in that area include the grape colaspis (*Colaspis brunnea*), fall armyworm (*Spodoptora frugiperela*), the rice stalk borer (*Chilo plejadellus*), sugarcane borer (*Diatraea saccharalis*), the leafhopper (*Draeculacephala portola*), the chinch bug (*Blissus leucopterus leucopterus*), and various species of shorthorn grasshoppers. *Stem borers* were important pests of rice in the South in earlier years; they seem to be increasing in importance. Insects that attack rice in California include the rice water weevil and the dipterous rice leaf miner (*Hydrellia grisceola*). The tadpole shrimp (*Triops longicaudatus*), although not an insect, sometimes causes damage to California rice.

QUALITY CHARACTERS

The milling, cooking, and processing characteristics are significant in rice in the United States. These characteristics are important considerations in breeding programs and govern the market demand of the rice produced by farmers. The basic characteristics of short-, medium-, and long-grain varieties are firmly established. Market demand is based upon these characteristics, so new varieties developed and released to farmers must conform to the established characteristics for each grain type. This has tended to narrow the genetic base of rice varieties in the United States. All long grains have essentially the same vulnerability to diseases, while the vulnerability of the four medium-grain varieties now available differs to some extent, even though they are related.

GENETIC UNIFORMITY

Although the crop is grown in localized areas, the varieties have relatively wide adaptation and acceptance. The rice industry is too limited to justify utilization of more than a few varieties at one time, but standby varieties with differing patterns of resistance might be developed and maintained for emergencies. The development of heterogeneous varieties as a hedge should be considered.

Only about a dozen varieties are reported to be grown in the United States. Fourteen major varieties are listed in Table 1. There may also be a few fields of R-D and Vegold. Over 90 percent of the southern rice acreage is planted to five varieties, and virtually all of the California rice acreage is planted to three varieties. Thus, the genetic base for commercial rice varieties grown in the United States is narrow.

There are only two short-grain varieties. Caloro was selected from a variety introduced from Japan. Colusa was selected from a Chinese variety obtained from Italy.

Four medium-grain varieties are grown by farmers in the southern area and each of these four varieties has Blue Rose in its pedigree. The three most important varieties also have Fortuna and two have Colusa in their pedigrees. Caloro was the recurring parent in the development of the medium-grain variety Calrose, so nearly 85 percent of the California crop has the same genetic base for disease response. Parental varieties are listed in Table 2.

TABLE 1   Percentage of the Rice Acreage in the United States That Is Cropped to Each Variety, 1970

Variety and Type	Arkansas	Louisiana	Mississippi	Texas	Southern States	California	U.S. Total
Short-grain							
Caloro^a	0.97	—	—	—	0.30	30.00	5.85
Colusa^a	—	—	—	—	—	16.77	2.92
Medium-grain							
Arkrose	0.57	—	—	—	0.17	—	0.10
Calrose^b	—	—	—	—	—	53.23	9.74
Nato	16.06	17.14	—	16.50	16.08	—	13.15
Nova	5.97	0.77	0.33	—	2.05	—	1.67
Saturn	1.27	56.90	—	0.59	20.60	—	16.86
Long-grain							
Belle Patna	0.99	2.92	0.27	31.84	11.38	—	9.30
Bluebelle	7.27	4.08	12.12	44.99	18.21	—	14.89
Bluebonnet	5.57	3.69	13.48	—	3.41	—	2.79
Dawn	0.93	0.82	0.69	4.12	1.88	—	1.54
Rexoro^c	—	—	—	1.03	0.33	—	0.26
Starbonnet	60.40	13.13	73.11	0.93	25.40	—	20.77
Toro	—	0.55	—	—	0.19	—	0.16

Source: Rice Millers Assoc. acreage report.

^{a} Short-grain varieties not identified by variety in the acreage report.
^{b} Includes small acreage of Earlirose and Kokaho Rose.
^{c} Includes TP 49.

All long-grain varieties grown on farms in the United States have Rexoro and Fortuna in their pedigree. These two varieties have the cooking and processing characteristics desired in long-grain varieties. This fact probably accounts for their universal appearance in the pedigrees of our long-grain varieties.

The United States has a large number of varieties in its world collection. Quite a few of these are being used in breeding programs to increase the diversity of material, including varieties from Japan, Taiwan, Ceylon, Philippines, India, Pakistan, Surinam, Italy, and Hungary. Varieties from other countries are tested to determine their possible usefulness in the breeding program. However, new germ plasm being introduced to convert to the semidwarf plant type may on the one hand reduce vulnerability to *Pyricularia oryzae,* and on the other increase susceptibility to such other diseases and disorders as straighthead and *H. oryzae.*

VARIETY-RELATED DISEASES

In a few cases rice diseases have caused widespread and catastrophic losses in yield in the United States. An epidemic of the blast disease in South Carolina early in this century may well have contributed to

TABLE 2   Parent Varieties of United States Rice Varieties in Commercial Production in 1970

		Present	Used as parent, Number of Varieties	
Variety	Grain Type	Commercial Variety	Medium-grain	Long-grain
Blue Rose	M	No	4	4
Caloro	S	Yes	2	0
Colusa	S	Yes	2	0
Edith	L	No	1	0
Fortuna	L	No	3	6
Hill long-grain	L	No	0	2
Lady Wright	L	No	1	0
Nira	L	No	1	0
Rexoro	L	Yes	1	6
Shoemed	M	No	2	1
Zenith	M	No	1	0
Unknown	—	No	0	2
C.I. 5094^a				
C.I. 5309^a				
C.I. 7689^a				
Carolina Gold^a				

^{a} These varieties in pedigree of Dawn.

the discontinuance of rice growing in the Southeast. White tip pro-duced severe losses in Arkansas from about 1935 to 1945. This dis-ease contributed to the reduction in acreage of Blue Rose and Early Prolific. At the present time there are varieties that are resistant to predominant races of *P. oryzae,* but races that are parasitic on cur-rent varieties have been identified. Early seeding reduces exposure to blast. Other varieties are resistant to white tip, and there are effective control measures for the causal nematode. Of the other diseases now occurring in the United States, brown leaf spot and stackburn, with associated damage to seed development, and bordered sheath spot seem the most likely to become widespread and devastating. Hoja blanca could also cause significant losses in yield, but the known vec-tor of this disease apparently does not survive well in the temperate zone. If a hardy biotype of the vector should arise or if an indigenous insect can transmit hoja blanca, this disease could be important in the United States. The other diseases now occurring in the United States cause small but nonetheless economic losses in yield and quality. However, it is thought unlikely that they will ever reach epidemic proportions. Seed rot and seedling diseases usually are more severe in rice sown early when the temperature of the soil or water is low. Un-decayed plant residue serves as a cultural medium for seed rot or-ganisms, therefore better seedbed preparation probably would reduce the incidence of the diseases.

Two major groups of diseases of rice do not occur in the United States. These are the bacterial and virus diseases that have caused widespread losses in many areas in Asia. Bacterial leaf blight caused by *Xanthomonas oryzae* (Uyeda and Ishiyama) Dows, and bacterial leaf streak, caused by *X. translucens* (sometimes referred to as f. sp. *oryzae*), do not occur in the United States, although they are major diseases in many areas in Southeast Asia. The United States varieties Early Prolific, Nova, and Zenith are resistant to bacterial blight. Of these three only Nova was grown commercially in the United States in 1970. It would appear possible to breed resistant varieties of other types, because Nova undoubtedly got its resistance from Zenith. Blue Rose is the only named United States variety reported to be resistant to bacterial leaf streak although other varieties in our collection of varieties are resistant.

In addition to the insect-transmitted virus diseases, there is one manually transmitted virus affecting rice. A strain of sugarcane mosaic virus was transmitted manually to five United States rice varieties in Louisiana. This disease is not found on rice growing in proximity to diseased sugarcane fields in Louisiana, so it is inferred that the virus is not a serious threat to rice.

The rice water weevil and the rice stink bug cause yield and quality losses to rice annually. There is an indication of differential varietal response to the rice water weevil and further research is under way in this area. Chemical control methods are available for the rice water weevil and the stink bug. The introduction of new insects is a potential threat to United States rice production because several insects that are extremely damaging to rice, for example, the vectors of Asian virus diseases, do not yet occur in the United States. If any of these insects were to be introduced into this country, the disease they transmit would be apt to occur soon afterwards. The rice gall midge (*Pachydiplosis oryzae*) is a serious pest of rice in many Asian countries, but it does not occur in the United States. The cereal leaf beetle (*Oulema melanopus*) occurs in the United States but not in areas where rice is grown. This insect will feed on rice but it requires dry soil in which to pupate, so it is unlikely to become a serious pest of flood-irrigated rice. *O. oryzae* is a pest of rice in Japan but not in the United States.

DISEASES AND PESTS IN THE TROPICS

The tropical climate is conducive to the proliferation of parasites, many of which take a serious toll from rice production. The important fungal diseases are blast, sheath blight (caused by *Corticium sasakii*), stem rot, and brown leaf spot. Blast is the most serious, occurring around the world. The fungus is highly variable, and many different races occur in the rice-growing areas. The fungus can attack the rice plant at all stages of growth from seedling to flowering, causing serious yield losses and even severe epidemics. Many cases of large-scale damage by this parasite have been reported from several countries, for example several thousand hectares of rice were destroyed by blast in the Philippines in 1969.

Sheath blight and stem rot cause damage in certain seasons. In addition to causing direct yield reduction, these two diseases reduce the straw strength and make the crop more prone to falling over. Brown leaf spot and narrow brown leaf spot are minor diseases in the tropics. The Bengal famine of 1943 caused by large-scale crop failure has been widely attributed to the attack of *Helminthosporium oryzae.* However, many plant pathologists are of the opinion that *Helmintho-*

sporium by itself does not cause serious damage. Serious damage is generally associated with such other primary problems as nutritional and physiological disorders. To support this view, pathologists cite the fact that the recent damage caused by *Helminthosporium* in the Ngale area of Indonesia and in Bohol island of the Philippines occurred on potash-deficient soils. Under good management the danger of large-scale attack of this disease appears minimal.

The two bacterial diseases of rice in tropical Asia, bacterial leaf blight and bacterial leaf streak, are widespread. The former is more destructive than the latter for it can attack the rice crop at all the stages of growth. A serious attack at the seedling stage, known as kresek, may kill the entire crop, and kresek attacks have been reported to have caused heavy losses in Indonesia and East Pakistan. The better-known symptom of the disease is blighting of the leaves. This occurs most often after the heads appear and cause serious yield reductions. Bacterial leaf streak is not a very serious disease. Its attack occurs in the rainy season generally after the typhoons and heavy winds that not only disseminate the bacteria but wound the leaves so as to provide ready entry. However, subsequent growth is free from the streak organism and the plant recovers rapidly. The bacterial diseases are unknown in Latin America.

Four virus diseases, tungro, grassy stunt, yellow dwarf, and orange leaf, are of widespread occurrence in tropical Asia. Of these, tungro (also known as yellow orange leaf in Thailand, Penjakit Merah in Malaysia, Mentek in Indonesia and yellowing in India) is by far the most important. This virus, transmitted by the green leafhopper, *Nephotettix impecticeps* has been reported from most of the major rice-growing countries of tropical Asia. Serious epidemics of the disease have occurred in Indonesia, Malaysia, India, Thailand, East Pakistan, and the Philippines and several thousand hectares of the crop were discarded after an epidemic in Malaysia in 1964. According to one report about one-third of the rice in Thailand is infected to varying degrees by the virus every year. Epidemics of this virus were reported in India in the West Bengal and Bihar states in 1970, and in recent years it has become a major problem in the Philippines where several hundred thousand hectares of the crop were badly affected in 1970, and again in 1971.

The grassy stunt virus disease is not as widespread but is potentially very dangerous. Besides the Philippines, its occurrence has been reported from India, Ceylon, and Thailand. In 1970 serious outbreaks occurred in Bacolod and Cotabato, two widely separated areas of the

Philippines. With the development of irrigation facilities and the introduction of continuous cropping of rice the incidence of this disease is likely to increase. The virus is transmitted by the brown planthopper, *Nilaparvata lugens.*

The yellow dwarf and orange leaf viruses are of minor importance, and it is unlikely that these will ever become serious diseases in the tropics mainly because, in the case of yellow dwarf, the incubation period in the vectors and the plant is more than a month. This disease, however, could become a serious problem on the ratoon crop, if large scale ratooning is introduced in Asia. Orange leaf is what may be termed a self-eliminating disease; the infected plants do not live very long, thus limiting the source of inoculum.

Hoja blanca has been a devastating virus disease of rice in parts of Latin America. Recent work, however, in developing varieties resistant to the virus and others resistant to the insect vector should greatly reduce losses from this virus.

Many different insect species are pests of rice in the tropics. Of these, the brown planthopper, green leafhopper, *Sogatodes orizicola,* gall midge, and the stem borers are of major importance. The first three cause serious damage by direct feeding and sometimes, due to heavy build-up of populations the plants are killed completely, a condition known as hopperburn. In addition, these three insect species transmit virus diseases. The stem borer larvae attack the young tillers of rice as well as the flowering shoots and take a heavy toll of rice in certain seasons. The gall midge is a serious problem in Ceylon, South India, East Pakistan, Thailand, Vietnam, and Indonesia.

CROP PROTECTION

Rice has traditionally been grown in the tropics without protection against pests and diseases. Limited efforts have been made in a few countries, particularly in India, to develop blast-resistant varieties, but no serious work has been done on insect-resistant varieties. Nor has much work been carried out on the chemical control of rice diseases in the tropics. The chemical control of high populations of parasites for prolonged periods under the monsoon climate is difficult. Moreover, the economic and social conditions in the tropics present serious obstacles to chemical control of rice diseases. Therefore, at IRRI, plant breeders and plant pathologists have emphasized breeding rice for disease resistance. Many other national rice improvement programs in the tropics have followed suit. Research on the chemical control of rice insect pests was initiated at the start of the Institute, and several systemic chemicals were found to be effective against common insect pests and are being used in much of tropical Asia. The recent discovery of rice varieties with a very high level of resistance to important insect species has opened up new vistas for the control of insect populations through the development of insect-resistant varieties. Therefore, breeding for disease and insect resistance is one of the major objectives of several breeding programs.

SOURCES OF DISEASE AND INSECT RESISTANCE

Plant pathologists and entomologists at IRRI have screened the world collection and have identified several varieties resistant to a particular disease or a pest. At least a dozen different varieties have been found to be resistant to blast in many countries. Some of them, such as Colombia 1, Tetep, Carreon, and Mamoriaka, have shown a very broad spectrum of resistance and are being employed in the breeding programs. Although the resistance to blast in most varieties appears to be specific, evidence suggests that the resistance in Tetep may be general.

Several dozen varieties with resistance to bacterial leaf blight have been identified, and some of them (like BJ 1) appear to have a broad spectrum of resistance. The resistant varieties come from different geographical areas and probably differ in their genes for resistance. The resistance to bacterial leaf blight in several varieties investigated to date appears to be governed by one or two major genes whose action is influenced by modifiers. Similarly, several varieties resistant to bacterial leaf streak, a polygenic characteristic, are available.

Varieties such as Peta, Pankhari 203, Sigadis, and Bengawan are resistant to the tungro virus and its vector, the green leafhopper. The virus resistance in these varieties is polygenic. Until recently no source of resistance to grassy stunt virus was available. A careful search for resistant varieties in the world collection yielded negative results. Many accessions of the wild species of genus *Oryza* were then screened. One accession of *O. nivara* was found to be highly resistant, its resistance being conditioned by a single dominant gene. *O. nivara* has a poor plant type, but it is compatible with *O. sativa* and no problem was encountered in transferring the gene for grassy stunt resistance to high-yielding lines by backcrossing.

More than 20 different varieties resistant to the green leafhopper are available. The resistance is under monogenic control; three independent loci for resistance have been identified; a dominant allele at any one of these loci conveys resistance. The Pankhari 203 gene and the ASD-7 gene give a higher level of resistance than the Peta gene.

At least 20 different varieties resistant to the brown planthopper are available. These varieties are homozygous for either of the two genes for resistance, one of which is dominant and the other recessive. The varieties Mudgo, MTU 15, and Co 22 possess a dominant gene for resistance, whereas the resistance of ASD 7 and H 105 is governed by a recessive gene.

Many varieties are resistant to direct damage caused by *Sogatodes orizicola.* Resistance, which is highly heritable, is probably monogenic and is easily combined with other desirable traits.

Several varieties resistant to stem borers, (TKM-6, Ptb 18, and EK 1263) are known, and resistance appears to be polygenic. Similarly, varieties with a high level of resistance to gall midge (EK 1263, Leuang 152, and Siam 29) are known.

BREEDING FOR RESISTANCE

Major emphasis has recently been placed on breeding for disease and insect resistance. Most of the varieties identified as sources of resistance described in the previous section have poor plant type and low yield potential. Therefore, these varieties were crossed with improved-plant-type, high-yielding varieties or experimental lines (Taichung Native 1, IR8, or IR262–43–8, and other dwarf selections). By now important sources of resistance to the major diseases and pests have been transferred to the improved plant type. Some of these lines were considered promising enough to be named as varieties. Thus, IR20 is resistant to bacterial leaf blight, and tungro and has a moderate level of resistance to stem borers. Resistant to bacterial leaf blight is IR22, IR24 is resistant to the green leafhopper and bacterial leaf blight, and CICA 4 is resistant to *Sogatodes orizicola.* However, these varieties are susceptible to other disease and pests. From the intercrosses of these new varieties as well as other resistant selections, the progenies obtained retain the desirable traits of the named varieties and in addition combine resistance to various diseases and pests not present in these varieties.

Promising selections at IRRI now being evaluated may combine resistance to green leafhoppers, brown planthoppers, and grassy stunt as well as the blast resistance of Tetep and the bacterial leaf blight resistance of IR22. The gall midge and stem borer resistance of EK 1263 are being transferred into these lines. It is hoped that within the next three to five years several varieties differing in growth duration and grain-quality characteristics but having resistance to the major diseases and pests will become available. Some of these will have different genes for resistance for the same disease or pest. This diversity will provide the defense mechanisms in case any of the insects and diseases develop new races.

Several national rice-breeding programs in the tropics have also embarked upon projects aimed at developing disease and insect-resistant varieties. The All India Coordinated Rice Improvement Project and the Breeding Division of the Rice Department in Thailand are working successfully, and the programs in Ceylon, East Pakistan, and Indonesia are paying increased attention to resistance breeding. The CIAT program in Colombia is emphasizing general resistance to blast. It is hoped that within the next few years several disease- and insect-resistant varieties will be available for the farmers throughout the tropics.

GENETIC VULNERABILITY OF HIGH-YIELDING VARITIES

A few of the tall, traditional varieties of Asia are resistant to one or more diseases or insect pests, but the great majority of them are susceptible to several. However, large-scale crop failures due to disease and insect attacks have not occurred in the recorded history of the crop with the exception of the Bengal famine of 1943. One reason for this may be the multitude of varieties grown in all the major rice-growing areas of tropical Asia, a genetic diversity that may have served the useful purpose of suppressing the build-up of parasite and insect populations.

The introduction of dwarf, high-yielding, and day-length-insensitive varieties with improved plant type has affected this situation in several respects.

First, many of the tall, traditional varieties are being rapidly replaced by a few genotypes. IR8 and other dwarf varieties with almost identical plant types are now being grown on an estimated 20 to 25 million acres in the tropics. This process of what might be called "genetic suffocation" is likely to continue in coming years. The new dwarfs have been confined largely to areas where water is available and controlled. Upland, deep-water, and temperate areas, unaffected to date by the new production varieties, in the next decade will likely adopt the dwarfs with comparable increases in productivity. Consequently, it may be predicted that the present area planted to dwarfs will continue to expand at the expense of numerous traditional varieties.

Second, the high-yielding varieties seem to be more prone to the attack of certain diseases and insects, such as sheath blight and brown planthoppers. The improved plant type varieties with high tillering, dwarf stature, and many leaves per unit area seem to create an ideal environment for the rapid multiplication of leafhoppers and plant-hoppers, as well as the sheath blight organism. Consequently, several instances of hopper burn in the fields of dwarf varieties caused mainly by the brown planthopper have been recently reported from the Philippines, Indonesia, Vietnam, Korea, Thailand, East Pakistan, and India. Similarly, the incidence of sheath blight in high-yielding varieties has been observed to be higher than that in the tall varieties.

Third, with the availability of high yielding, nitrogen-responsive varieties, the farmers are increasingly using more fertilizer and are practicing more thorough weed control. The resultant luxuriant growth of the crop also favors the rapid multiplication of disease organisms and insect pests. Other things being equal, the danger of serious crop losses from the attacks of major diseases and insects is greater today than in the past. However, the improved cultural practices have reduced the vulnerability of the crop to organisms like *Helminthosporium* that are associated with nutritional or physiological disorders.

Fourth, with the development of day-length-insensitive varieties and irrigation facilities, the area under continuous cropping with rice has been increasing. Continuous cropping with susceptible varieties of rice increases the likelihood of large-scale build-up of insect populations and disease organisms, especially the viruses and their vectors. Recent outbreaks of the tungro and grassy stunt viruses in the Philippines have been attributed to large-scale introduction of double cropping with rice. Serious tungro infections were observed in the fields planted to vector-resistant but virus-susceptible varieties. On the other hand, very little virus damage was noticed in the fields planted to vector-susceptible but virus-resistant varieties. Similarly, a high proportion of plants infected with grassy stunt has been observed in the vector-resistant varieties and selections such as Mudgo and IR1561. Since the vectors are able to feed on the resistant plants for short periods, they can inoculate the plants. Field observations in the Philippines indicate that vector resistance alone is not enough for protecting the crop from tungro and grassy stunt. Therefore, IRRI breeders are emphasizing the breeding of virus-resistant varieties.

Since the vectors bring about serious yield losses by direct feeding, incorporation of resistance to the vectors is equally important. Moreover, the resistance to the tungro virus is under polygenic control and it will presumably be more stable than vector resistance, which is under monogenic control. In Latin America, where rice is directly seeded and not transplanted as in Asia, workers feel that resistance to the insect vector of hoja blanca will also control the virus disease. Thus, they presently emphasize breeding for insect resistance.

IS A COMMON GENE FOR DWARFISM DANGEROUS?

The relation between T cytoplasm and southern corn leaf blight raises question about the rice dwarfs that are now being cultivated on vast acreages in the tropics. Specifically, is there any danger of all the dwarf varieties succumbing to some disease or insect simply because they all have the same dwarfing gene? The consensus is that they will not. At present four IRRI-developed and named varieties are being grown in almost all the countries of tropical Asia. In addition at least 10 IRRI-developed dwarfs have been named as varieties in other countries and are being grown on large areas. Three dwarfs have been bred and released in Thailand, at least 12 in India, and 1 in Colombia. All of these dwarfs have either IR8 or TN1 or a related dwarf selection as one of the parents. However, the other parents of these varieties are quite unrelated and often have no common ancestry. Therefore, unless the dwarf gene itself or another gene nearby increases susceptibility to a particular pathogen or pest, it is improbable that all of them would inherit the common weakness. No evidence of strong linkage between the dwarfing gene and genes for resistance/susceptibility to any of the major diseases and insects has been found to date. Therefore, it appears unlikely that the common gene for dwarfing in all the new varieties would, *per se,* increase the genetic vulnerability to any disease or pest. Since many new dwarf varieties are rapidly being developed throughout the tropics, genetic diversity among the dwarfs will continue to increase.

SUMMARY AND CONCLUSIONS

● Cultivated rice is highly variable. Germ plasm has been collected, described, and is available to all rice workers. Collection from certain countries must be strengthened.

● The interdisciplinary approach at the international centers and in several national programs is increasing among all rice workers. Recognition of the interactions among host, parasite, and environment obliges breeders, pathologists, and entomologists to work jointly and offers the only rational approach toward steady progress.

• The international institutes have established a series of international disease and insect nurseries, where varieties are evaluated for reaction to diseases and insects. These identify sources of both broad and narrow resistance. They also indicate regional changes in parasites and pests and provide time to obtain resistance before widespread damage occurs.

• Rice breeders searching for general or field resistance should be cautioned against employing intercrossing of many lines followed by repeated cycles of unmodified bulk selection. Although this approach might accumulate many genes for resistance under parasite pressure, the interplant competition for light and nitrogen would eliminate the desired phenotypes.

• A relatively few highly productive dwarfs are replacing traditional varieties in the tropics. This trend is expected to continue with added reduction of varietal diversity.

• No danger associated with the single common dwarfing gene in the majority of the new varieties has occurred.

• The new plant type combined with double cropping, and increased fertilizer and weed control will undoubtedly aggravate certain diseases and pests. This is well recognized by rice workers. Incorporation of broad resistance for blast, grassy stunt, sheath blight, and certain planthoppers is being strengthened. These are the organisms most likely to assume greater importance in coming years.

• Other formerly minor diseases and pests may respond to the new host-microenvironment cultural practice complex and become major threats.

• Predictability for new threats is low. However, there has never been a greater number of well-trained people in mutual contact to handle problems as they arise. The tropical centers, with their linkages to national programs, offer competent manpower and the flexibility necessary to meet possible changes in parasites.

CHAPTER 12

Potato, Sugar Beet, and Sweet Potato

Contents

POTATO	191
History	191
Genetic Background	192
The Present Situation	193
Disease Resistance	194
Insect Resistance	196
Genetic Base of Major Varieties	197
Breeding Programs	199
Objectives; Approaches; Germ Plasm Resources; Current Breeding Programs	
Summary	202
SUGAR BEET	203
SWEET POTATO	204

190

POTATO

HISTORY

The cultivated potato (*Solanum tuberosum*), like many other crops, is not native to North America. The first recorded introduction was in Londonderry, N.H., in 1719 of tubers brought from Ireland. Among the later introductions the most important appears to have been the collection made in 1851, which Goodrich found was adapted to the long-day conditions in the State of New York. Called Rough Purple Chili, it came from a sample of tubers from a Panama market sent to Goodrich by the United States Consul; he assumed it had been sent from Chile. Since then it has been suggested that Rough Purple Chili was more likely to have been a long-day selection from Colombia.

In Europe the potato was introduced considerably earlier, probably about 1570 in Spain, and from there spread to continental Europe. During the period 1588–1593 it reached England from the New World. These introductions almost certainly belonged to the andigena group of *S. tuberosum* and were adapted to short-day conditions. They probably came from Colombia. In Europe, where short days are soon

191

followed by killing temperatures, it was possible to grow tuber crops only in warmer climates such as Spain or the west of Ireland. It was not until about the middle of the eighteenth century (Hawkes, 1967), when selection for early maturing or long-day forms was successful, that the potato began to be cultivated throughout Europe.

The importance of Goodrich's work is illustrated by today's leading variety in North America, Russet Burbank, which was derived from Rough Purple Chili. In 1857 Garnet Chili, an open-pollinated seedling of Rough Purple Chili, was released by Goodrich. In 1861 a breeder named Breese released Early Rose, a seedling of Garnet Chili. Luther Burbank's seedling was released in Massachusetts in 1873, one of 23 seedlings of Early Rose that he grew. The origin of the russet mutation that gave rise to Russet Burbank (or Netted Gem) is not known.

Many other varieties widely grown in North America and elsewhere originated from this material.

GENETIC BACKGROUND

Solanum tuberosum is a tetraploid (2n = 4x = 48), that is, each cell contains four sets of chromosomes. There are two groups within the species, which some consider to be subspecies. The tuberosum group includes the domestic potato, which also grows on the coastal plains of Chile, whereas the andigena group is confined to high altitudes in the Andes. In the tuberosum group are long-day plants forming tubers in the higher latitudes north and south of the equator in 18-hr days. In the andigena group are short-day plants forming tubers only in 12-hr days. Whether Goodrich's Rough Purple Chili was a tuberosum or a long-day andigena remains uncertain.

The crop is propagated vegetatively by planting certified tubers. Thus all the plants of one variety are, barring mutation, identical, and constitute a clone. The varieties are not true breeding (i.e., they are heterozygous), and so their seedling progenies are varied. Conventional breeding programs have repeated crosses among varieties or lines that have already yielded successful varieties. Large seedling progenies are then tested to select the very few forms that might exceed the parent varieties in yield, quality, and disease resistance.

Although there is at present no shortage of cultivated forms of andigena and more than 50 wild tuber-bearing species that can readily be crossed with tuberosum, new germ plasm has had relatively little impact on the development of present-day popular varieties. So far most new varieties with new germ plasm have not successfully competed with the established varieties and their derivatives, whose genetic base is narrow.

THE PRESENT SITUATION

Accurate statistics on the relative importance of current potato varieties are readily available because records must be kept showing that all seed potatoes shipped across state boundaries are certified as free of certain diseases and true to variety type. Most of the certified seed is grown in those northern states and parts of Canada where the cool summers limit aphid populations, thus reducing the risk of virus infection, and promote optimum tuber maturation. In recent years the annual total acreages of each variety certified by each seed-producing state and Canada have been summarized and published in the *Potato Handbook* of the Potato Association of America or in the *American Potato Journal.* We have assumed that these acreages are a reliable guide to the relative importance of the varieties. In 1969 the total United States certified seed acreage was approximately one-seventh of the overall potato acreage. Because most growers find that it pays to plant certified seed, the varietal composition of certified seed should closely approximate that for total production. In the seed-producing areas those acreages that fail certification are sold as table stock or for processing and are not recorded. We have, again, assumed that this applies to all the varieties in proportion to their acreage. A possible source of error arises from the fact that at least one large processor conducts his own breeding program and seed certification scheme, supplying tubers under number designations to contracting farmers.

This report concentrates on the pattern for North America as a whole. The varieties a state grows for certification are not necessarily the same as those grown for table stock or processing. Bulking the data overcomes these discrepancies, since the traffic in certified seed tubers into or out of the North American continent is small compared with internal production and use.

Table 1 shows the varieties that made up 1 percent or more of the total acreage in 1969 and 1970. The 1969 data are from Turnquist (1970); the 1970 data were obtained directly from the certification agencies. In 1969 92 varieties were certified (82 in 1970). In both years the four leading varieties made up approximately 72 percent of the acreage. The lower part of Table 1 shows the more important

TABLE 1   Varieties Constituting 1 Percent or more of 1969 Certified Acreage

	Percent of acreage	
Variety	1969^a	1970^b
Russet Burbank (Netted Gem)	25.0	28.1
Kennebec	18.6	20.0
Katahdin	18.4	15.3
Sebago	10.0	8.8
Norgold Russet	3.7	2.7
Red Pontiac	3.3	2.7
Norland	3.2	3.2
Norchip	2.3	4.4
Superior	2.3	2.0
Red La Soda	2.1	1.3
Irish Cobbler	1.5	1.0
La Rouge	1.5	0.8
	91.9	90.3
Other varieties with 900 or more certified acres in 1969		
White Rose	.9	.9
Chippewa	.8	.7
Green Mountain	.7	.6
Pungo	.7	.6
La Chipper	.5	$.3^c$
Red McClure	.4	$.3^c$
Norchief	.3	$.3^c$
	96.2	94.0

Source: Turnquist (1970) and certification agencies.

a 1969 acreage – 258,064.
b 1970 acreage – 267,000.
c Less than 900 acres.

minor varieties having certified acreages in 1969 of 900 or more. Some of these are locally important. This group could also include varieties that will become more popular.

DISEASE RESISTANCE

The reactions of the major varieties (shown in Table 1) to the more important diseases are shown in Table 2.

Breeding for late blight resistance has now switched from race-specific resistance to uniform resistance. Kennebec, which made up 20 percent of the 1970 acreage of certified seed, and Pungo, which made up 0.6 percent, both carry the gene $R1$ for blight resistance. Kennebec, which also has a low level of field resistance to late blight,

TABLE 2 Resistance to Potato Diseases

Variety	Late blight, *Phytophthora infestans*	Common scab, *Streptomyces scabies*	*Fusarium* tuber rot, *F.* spp.	Blackwart, *Synchytrium endobioticum*	Silver scurf, *Spondylocladium atrovirens*	*Verticillium* wilt, *V. albo-atrum*	Bacterial wilt, *Pseudomonas solanacearum*	Leafroll and Net necrosis	Yellow dwarf	Mild mosaic	Rugose mosaic	Corky ringspot
Russet Burbank		+							+			
Kennebec	+		+	+				+		+	+	
Katahdin				+		+	+	+		+	+	
Sebago	+	+				+	+	+	+	+	+	
Norchip		+			+							
Norland		+										
Red Pontiac												
Norgold Russet		+										
Superior		+									+	
Pungo	+	+						+		+	+	
Red LaSoda										+		+
Irish Cobbler		+	+	+						+		
La Rouge		+										
White Rose												+
Chippewa						+	+	+		+	+	
La Chipper	+											
Green Mountain				+			+					
Red McClure				+								
Norchief		+										

+ Indicates the variety carries resistance to disease but does not indicate the level or type of resistance.

195

is popular for other reasons. High levels of field resistance are available in forms such as the Mexican variety Atzimba, which can be grown successfully in areas like the Toluca Valley of Mexico where the blight hazard is great. Sebago and La Chipper carry low levels of field resistance. The leading North American variety Russet Burbank is also one of the most susceptible to late blight. There is much to recommend continued breeding for field resistance to late blight, the most important advantage being to reduce or eliminate both the cost of applying protective chemicals and their hazard to the environment.

Resistance to *Verticillium* wilt is available and present in several of the varieties. A current problem is the spread of this disease by seed tubers of Kennebec, and several other important commercial varieties that are susceptible to *Verticillium*. Symptoms of *Verticillium* wilt commonly are not evident until late in the season at the onset of maturation. The last field inspection is often made before symptoms appear. Removal of vines preparatory to harvest makes detection of wilt in the field impossible. It is difficult to detect the mycelium in the skin of infected tubers at harvest, but quick laboratory methods are available. Amendment of seed certification regulations for susceptible varieties to require a late season vine inspection and laboratory testing of tuber samples appear to be effective remedies. A shift from susceptible varieties similar to Kennebec, Irish Cobbler, Norchip, and Russet Burbank to such resistant varieties as Abnaki, Katahdin, Mohawk, and Ona for several seasons may well reduce the inoculum potential of *Verticillium* in the soil.

INSECT RESISTANCE

In general the development of varieties resistant to insects has been neglected. Although hopperburn resistance has been reported in Katahdin and Sebago (*Yearbook of Agriculture,* 1953) and fleabeetle resistance in Norchip and Norchief, the levels of resistance are not enough to withstand an onslaught of these pests. Effective levels of resistance to the leafhoppers, the potato fleabeetle, the Colorado potato beetle, aphids, and almost all potato insects, are present in the tuber-bearing *Solanum* species.

The changes in potato insect populations from 1956 to 1965 have been recorded by Landis *et al.* (1967). In the Red River Valley a decline in the Colorado potato beetle has been recorded, but an outbreak of the six-spotted leafhopper occurred in 1957. In Florida resistance of wireworms to chlorinated hydrocarbons has occurred, and leaf miners were prevalent in the early 1950's because they had become resistant to parathion; in this case the leaf miners were controlled by other insecticides. Cutworms have been troublesome in the Connecticut Valley, whereas the Colorado potato beetle has been the most important pest on Long Island during the decade. In Maine potato-infesting aphids have continued to be the most important insect, but the Colorado potato beetle caused some concern toward the end of the 10-yr period. In the Pacific Northwest aphids continued as a major insect but the populations varied considerably from year to year and place to place. The two-spotted mite caused difficulty where DDT had been used heavily, and the garden symphylan has caused damage to tubers. Wireworms of one kind or another continue to cause damage in the Pacific Northwest.

GENETIC BASE OF MAJOR VARIETIES

The pedigrees of most varieties are published. Figure 1 illustrates the relationships between 12 of the 19 varieties listed in Table 1. Two of the older varieties (Russet Burbank and Irish Cobbler) are featured as parents in these pedigrees. Inspection of the data shows that some varieties are sibs (Katahdin and Chippewa), others half-sibs (Sebago and Red Pontiac, Norchief and Norland, Kennebec and Superior) and that backcrosses are not uncommon. If we bear in mind the fairly recent origins of modern potato varieties and that they are, for the most part, derived from the survivors of the late blight epidemics of the 1840's in Europe and North America, it seems likely that the genetic base was already somewhat narrow by the time modern potato breeding got under way. The fivefold increase in yield resulting from selection during the last 100 years of potato improvement has produced a group of varieties that are genetically similar and unlikely to respond to further selection for yield. In the long run response to selection for other characters is also likely to be limited. The introduction of new genetic material has been limited to the major genes for disease resistance, such as the *R* genes for late blight resistance from *Solanum demissum.* Until recent years no major introduction of genetic variability has been made.

Examination of the pedigrees shows that no cytoplasmic uniformity results from consistently using only a few ancestral varieties as female parents.

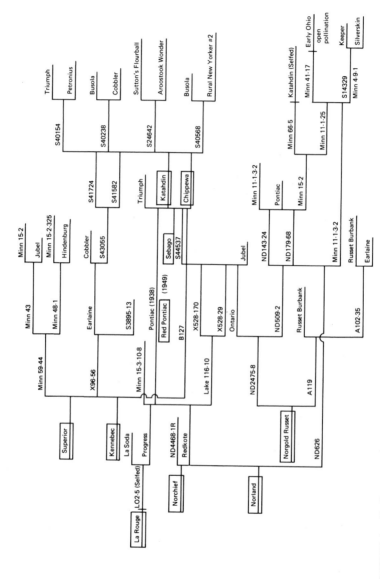

FIGURE 1 Pedigree to show origins and relationships among 10 (in boxes) of the widely grown potato varieties listed in Table 1. Inputs from Russet Burbank and (Irish) Cobbler are also shown.

BREEDING PROGRAMS

Objectives One of the most important factors regulating the potato varieties grown in North America has been the increased use of potatoes for processing. This has created a demand for varieties tailored for particular uses. No single variety is satisfactory for all purposes. That Russet Burbank comes as close as any no doubt accounts for its popularity. Top-grade tubers of Russet Burbank command a premium price as table stock for baking, and the lower grades find a ready market among processors.

The increasing importance of processed potatoes has resulted in a trend away from red-skinned varieties to white varieties. Processors have for many years produced chips and dehydrated or "instant" potato, and they now manufacture extruded potato for processing into granules, nuggets, and chips. The products require a potato with a high starch content and, like tubers for potato chips, a low content of reducing sugars so that frying does not produce too dark a color in the final product. New varieties are needed for processing into french fries that do not become soggy after freezing and reheating. Some canning companies require translucent tuber flesh for processing in potato soup. There are also stringent requirements for tubers that do not deteriorate in storage.

The market requirements for tubers will no doubt become even more specialized and exacting than they are now. One major processor has a long-established breeding program with variety trials and seed tuber production to ensure that its contracting farmers produce materials to meet its needs.

Approaches Several approaches to breeding new potato varieties are in use. The most important and most widely practiced is the traditional approach, but two other methods will also be discussed.

The traditional approach is pedigree selection, which uses old varieties or breeding clones that have given rise to good varieties in the past, to obtain a combination of desirable qualities in one clone. Success depends on growing very large numbers of seedlings and on accurate scoring. The parent lines must be sufficiently fertile to ensure adequate numbers of seedlings. This restriction has further narrowed the genetic base since some otherwise desirable parents have low fertility and do not produce enough seedlings.

Breeding programs that have emphasized disease resistance have drawn on the extensive collections of wild South American tuberous species by crossing them with breeding lines. The extraction of single dominant genes has been the chief objective. Varieties and breeding lines resistant to late blight, virus diseases, nematodes, and several insect pests have been produced in this way (Rowe, 1969).

The importance of uniform resistance to late blight has alerted breeders to the advantages of similar resistance to such other pests as the golden nematode, which, like late blight, is not always satisfactorily controlled by major gene resistance alone.

The traditional breeding approach has been successful and continues to be widely used. It is unlikely, however, to show a continued increase in yield.

Simmonds several years ago advocated a new approach to potato breeding through intensive selection for increased tuber size and yield under long-day conditions in very mixed populations of andigena material. The procedure employs alternating cycles of tuber- and seed-propagation. This approach is analogous to bulk population breeding in other crops, a method that appears to be useful and promising. Breeding lines derived from Simmonds' andigena material are now being used, together with a very wide variety of andigena introductions from South America by Plaisted at Cornell.

A second method makes use of haploids of *S. tuberosum* produced by the method developed by Hougas and Peloquin (1958). Haploids, that is, plants that have half (24) the normal number of chromosomes, of the common potato have been obtained in large numbers from a wide range of parent varieties. They have been hybridized with many cultivated and wild 24-chromosome tuber-bearing *Solanum* species from Mexico and South America. The progenies are variable, but many lines are vigorous, fertile, and high yielding. These hybrids with 24 chromosomes are the initial tools for simplified genetic studies with potatoes. They also provide genotypes with much genetic diversity that can be used for potato improvement.

Many lines with 24 chromosomes have been hybridized with commercial varieties having 48 chromosomes. A surprising result is the high yield of the 48-chromosome progeny from these crosses. The mean yields of several families exceeded the yield of the best potato varieties. The basis of the high yield appears to be the method by which the haploids form pollen grains with 24 chromosomes. The chromosomes fail to separate at the first division leading to formation of the pollen grains (meiosis), which means that a large proportion of the genes that were heterozygous remain heterozygous in the pollen that is formed. It is therefore a new method of transferring useful gene combinations from parent to offspring. It is also an effective method of germ plasm transfer from wild potato species to cultivated forms.

Germ Plasm Resources Several countries, including the United States, maintain collections of *Solanum* germ plasm. These collections consist of clones or seed stocks of wild species, hybrids, breeding lines, and commercial varieties. In a recent survey of three of the largest collections (Commonwealth Potato Collection, East German Collection, and the Collection at the Potato Introduction Station, Wisconsin), Rowe (1970) noted that of 93 wild species being maintained, 40 were represented by only 1 to 4 accessions and 17 of these were represented by only 1 living specimen. Seven species accounted for more than 50 percent of the total accessions. These latter were the species that have been most widely used in breeding. Rowe concluded that broader collections are needed and that they should be established in the centers of *Solanum* diversity by further developing collections already established in Mexico, Colombia, Peru, and Argentina. The existing collections of varieties appear to be adequate, probably because, except for virus deterioration, they are easy to maintain.

Those responsible for some collections, such as the interregional one in Wisconsin, have taken the approach that the bulk of their material should be available to the breeder in a usable form. Material is maintained in populations rather than clones, as far as possible in a form that can be planted and harvested by machine.

Current Breeding Programs The most extensive potato breeding program in the United States is conducted by the USDA. Its goal is the development of multiple pest-resistant, high-yielding varieties adapted to specific areas and uses. The breeding work is carried out cooperatively with Maine, Idaho, Washington, and Colorado, and trials in 22 states and 8 foreign countries ensure adequate evaluation. Six of the 19 varieties listed in Table 1 are products of the USDA program. While much of the program is based on traditional methods, improved systems of potato breeding are sought. Other important features include emphasis on resistance to major insect pests, and polygenic resistance to late blight and golden nematode. Work is also under way to improve the protein content and other nutrient qualities in varieties adapted to potato growing regions in developing countries. In cooperation with all 50 states the USDA maintains the Inter-Regional Potato Collection at Sturgeon Bay, Wisconsin.

In addition to the collaborative program with the USDA, the following states have active potato breeding programs: Maine, North Carolina, New York, Louisiana, Michigan, Wisconsin, Minnesota, North Dakota, Nebraska, and Indiana. One major processor has a long-established breeding program, and another has recently begun. The Potato Research Station in New Brunswick services the needs of Canada.

Outside North America, the indications are that the genetic base of commercial potatoes is as narrow as here. There are major potato breeding programs in Poland, USSR, Mexico, Scotland, England, Holland, East and West Germany, Scandinavia, Japan, Australia, New Zealand, Kenya, and India. Much of the work of these centers is along conventional lines.

SUMMARY

Present-day commercial varieties are very closely related. Although the original genetic base 100 or more years ago was broad, it has narrowed. The means of broadening the base are available and are being used to an increasing extent.

Are there dangers from epidemic pests and diseases in the future that are inherent in our present technology? We think not, but we can ask the question: Are there undescribed diseases or insects now present in Central and South America that could be catastrophic if introduced into North America or Europe? As we increasingly use the species from these geographic areas as sources of resistance to the currently known pests and diseases of the temperate zone countries, we may be unknowingly breeding for susceptibility to an unknown *Cercospora* or *Septoria.* How well do we know the diseases now prevalent on the native *Solanum* species in Central and South America? The inherent dangers of spreading pests or diseases via seed tubers are controlled by the seed certification schemes. As new varieties are brought out to meet processing requirements, they may have special problems that are now nonexistent or minor, but varietal development and improvement should take care of such problems.

What catastrophes can we foresee? At the present time we rely on chemical protectants more heavily than we should. Breeding to relieve this situation is under way. If new varieties should result in dramatic improvements in yield we may be, in the end, slightly worse off, with perhaps one variety making up 72 percent of our crop instead of four not greatly different ones.

The answer appears to be development of a back-up capability. If our best varieties are subject to epidemics that result primarily from genetic uniformity and if economic considerations demand that we take this risk, then we must have material ready to fall back on. The role of the public breeder, both USDA and experiment station, is to preserve this flexibility.

SUGAR BEET

In 1970 the total United States sugar beet (*Beta vulgaris*) acreage was about 1.4 million. The four states with the largest plantings were California (305,000), Idaho (169,000), Minnesota (146,800), and Colorado (146,400). The crop, which is raised from seed each year, is grown under contract for the major sugar companies, who have built processing plants at strategic locations. Over 90 percent of the seed is hybrid and produced with the aid of one source of cytoplasmic male sterility originally discovered by Owen, USDA, Utah. Cytoplasmic male sterility has been found in several different open-pollinated varieties, but there is no evidence that these sources have significantly different characteristics.

Seeds of sugar beet differ from those of most other plants. The embryos are enclosed in the dry flower parts that make up a seed ball. These are called multigerm seeds. Male or pollen parents are usually multigerm. The technology of hybrid seed production also makes use of a gene for monogerm seed. That trait makes it unnecessary to thin following machine seeding. The female or seed parent is not only male sterile but monogerm. All hybrids utilize the same recessive gene for monogerm seed originally discovered by Savitsky (1950).

Two public hybrids, US H9 and US H20 (developed by USDA breeders), and private hybrids similar to H9 and H20 together account for about 40 percent of the total United States acreage. US H9 and US H20 do not have common parental lines. The remaining acreage is planted to cultivars developed by the sugar companies; five of the major companies have breeding programs, and their hybrids are grown in the Great Plains states and in Washington. The kinds of hybrids used, the extent to which they are grown, and their reactions to the major diseases are shown in Table 3.

The sugar beet originated in Central Europe. Since its introduction to North America selection for high sugar yield and disease resistance has narrowed the genetic base. Breeders are especially concerned

TABLE 3   Sugar Beet Hybrids in the United States, 1971, and Their Disease Reactions

	Type of Hybrid	1971 Acreage	Cercospora Leaf Spot	Aphanomyces Root Rot	Curly Top	Virus Yellows
US H9	3-way	290,000			R^d	R
US H20	3-way	300,000^b	R	R	R	
Amalgamated H-1	4-way	110,000			R	
Amalgamated H-3	4-way	40,000			R	
American #3 Hybrid A	3-way	20,000	R	R		
American #3 Hybrid N	3-way	30,000				
American #3 Hybrid T	3-way	25,000	R	R		
Great Western type A	4-way	228,000	R			
Great Western type B	3-way	15,000	R		R	
Great Western type C	3-way	12,000			R	
Great Western type D	3- and 4-way	43,000				
Great Western type E^a	3-way	30,000	R	R		
Holly H10		60,000	R		R	
Holly H19		40,000			R	
Utah-Idaho Hybrid 3^c	single X	62,000			R	R
Utah-Idaho Hybrid 8^c	single X	80,000			R	R

^{a} Uses same pollinator as US H20.
^{b} Includes 135,000 acres of a similar hybrid American #2B.
^{c} Use same monogerm cytoplasmic male-sterile parent and related pollinators.
^{d} Indicates resistance.

about the narrow base of resistance to curly top, a virus spread by the beet leafhopper. The genetic uniformity imposed by cytoplasmic sterility and the requirement for monogerm seed are of great immediate significance since dependence on such technological features creates just the kind of genetic vulnerability that this report is concerned with. While no one can point to the equivalent of a *Helminthosporium maydis* or an *H. victoriae* waiting in the wings, they may be there.

Sugar beet breeders, both public and private, are aware of this hazard and are at work to provide alternative male-sterile cytoplasms and to broaden the genetic base of disease resistance. It would also seem worthwhile to look for other genes for monogerm seed since the industry has come to depend on it.

Collections of wild species are maintained in the United States by the Sugarbeet Investigations unit of the USDA, Iowa State University, and several of the private breeders.

SWEET POTATO

The sweet potato (*Ipomoea batatas*) is vegetatively propagated from certified seed roots. Like the potato it is a polyploid, each cell having six sets of chromosomes. Also, the sweet potato is highly variable

TABLE 4 United States Sweet Potato Varieties in 1970 and Their Disease Reactions

Cultivar	Acreage	Per-cent	Root Knot	Soil Rot	Stem Rot	Black Rot	Scurf	Internal Cork	Sclerotial Blight
Centennial	95,000	68.8	VS	VS	MR	S	S	MR	S
Nemagold	11,000	7.9	R	MR	S	S	S	MR	
Georgia Red	10,000	7.2	S	S	S	S	S	S	
Porto Rico	4,000	2.9	S		S	S	S	S	
Red Velvet	4,000	2.9							
Jewel	3,000	2.2	MR		R			R	
Jersey types	6,000	4.3	MR	S	S	MR	S	R	
Others	5,000	3.9							
Total	138,000								

VS = very susceptible
S = susceptible
MR = moderately resistant
R = resistant

when bred from seed. In 1970 the total United States crop occupied
137,000 acres, the two largest producers being Louisiana (49,000)
and North Carolina (25,000). Eleven other states with acreages ranging from 1,700 (Arkansas) to 13,000 (Texas) made up the rest of the
1970 total.

The leading varieties, their acreages, and reactions to major diseases
are shown in Table 4. Some 48 varieties of sweet potato were developed and released in the United States during the period 1939 to
1969. These vary in their resistance to pests and diseases. The leading variety, Centennial, is extremely susceptible to root knot and soil
rot. Among virus diseases, only resistance to internal cork is available.
The root knot nematode (*Meloidogyne incognita*) can be controlled
by resistant varieties, but this resistance is race specific. Stem rot,
caused by *Fusarium oxysporum batatas,* can also be controlled by
resistant varieties. Although there is partial resistance to black rot,
caused by *Ceratocystis fimbriata,* control depends on using disease-
free plants, crop rotation, and sanitation. Of the other important diseases, resistance to soil rot, caused by *Streptomyces ipomoea,* appears
promising, whereas the resistance to scurf, caused by *Monilochaetes
infuscans,* may be of little value since it permits development of
small root lesions.

There are four major insect pests: sweet potato weevil (*Cylas
formicarius elegantulus*), wireworm (*Conoderus falli*), banded cucumber beetle (*Diabrotica balteata*) and sweet potato flea beetle
(*Systena elongata*). There is effective resistance to all except the first
of these. Breeding for insect resistance is increasing in importance,

particularly since several effective chemical controls have been prohibited.

Current breeding programs in Charleston, S.C., Raleigh, N.C., Baton Rouge, La., Tifton, Ga., College Park, Md., and State College, Miss., are concerned with increasing genetic diversity. Breeders are well supplied with material from the United States Plant Introduction Service Station at Glen Dale, Md. Research on sweet potatoes in the United States during the last 30 years was reviewed by Hernandez *et al.* (1971).

The major genetic hazard to sweet potatoes appears to be the predominance of one variety.

REFERENCES

Hawkes, J. G. 1967. The history of the potato. J. R. Hortic. Soc. 92:207–224; 249–262; 288–302.

Hernandez, T. P. [ed.]. 1970. Thirty years of cooperative sweet potato research. *In* South. Coop. Ser. Bull. 159, La. Agric. Exp. Sta.

Hougas, R. W., and S. Peloquin. 1958. The potential of potato haploids in breeding and genetic research. Am. Potato J. 35:701–707.

Landis, B. J., M. Semel, and G. W. Simpson. 1967. A decade of change in insect population on potatoes 1956–1965. Potato Handb. 12:12–29.

Rowe, P. R. 1969. Nature, distribution and use of diversity in the tuber-bearing *Solanum* species. Econ. Bot. 23:330–338.

Rowe, P. R. 1970. The dimensions of existing *Solanum* germ plasm collections. Am. Potato J. 47:205–208.

Savitsky, V. F. 1950. Monogerm sugar beets in the United States. Proc. Amer. Soc. Sugar Beet Tech. 156–159.

Turnquist, O. C. 1970. Production of certified seed potatoes by varieties–1969. Am. Potato J. 47:138–142.

CHAPTER 13

Soybeans and Other Edible Legumes

Contents

SOYBEANS	209
Origin and History	209
Importance of Crop	210
Genetic Uniformity	211
Pests and Diseases	214
Status of Breeding for Resistance	214
PEANUTS	217
Origin	217
Importance of Crop	217
Genetic Uniformity	218
Pests and Diseases	221
Status of Breeding	223
DRY BEANS	224
Origin, History, and Importance	224
Production Patterns and Epidemic Potential	224
Genetic Diversity	228
Strategy of Varietal Development	229
Genetic Control of Resistance to Specific Diseases	230
Common Mosaic; Anthracnose; Halo Bacterial Blight;	
Common Bacterial Blight; Bacterial Wilt; Root Rot;	
Rust; Nonparasitic Diseases	
Preserving a Gene or a Polygenic System	234

207

SNAP BEANS   234
  History and Importance   234
  Genetic Uniformity   236
  Market Requirements   238
  Diseases and Insect Pests   238
      Common Mosaic Virus; Curly Top; Yellow Mosaic
      Virus; Halo Bacterial Blight; Root Rots; Rust
  Sources of Germ Plasm   242
  Measures to Reduce Vulnerability   242
PEAS   244
  History   244
  Importance of Crop   245
  Genetic Diversity   246
  Diseases and Insect Pests   247
  Genetic Vulnerability   249
  Measures to Reduce Vulnerability   251

SOYBEANS

ORIGIN AND HISTORY

The soybean (*Glycine max*) is a native of eastern Asia. Generally accepted as its wild ancestor is *Glycine ussuriensis,* which is found throughout eastern Asia. Since its domestication soybean seed has been used for a variety of foods in the Orient. Seeds are also crushed and the oil used for food and food products. The cake that remains is used for animal feed and fertilizer. In the United States soybeans were first grown for hay and green manure, but are now grown for the beans, which are processed for oil and high-protein meal.

The first reported plantings of soybeans in the United States were in 1804 by Mease. Numerous trials were reported during the nineteenth century, most of which indicated that the soybean grows well under United States conditions.

According to Piper and Morse (1910), not more than eight varieties of soybeans were grown in the United States prior to 1898. By 1910 280 varieties or types were being grown, most of which had been obtained from China, Japan, and India. A few came from other countries in eastern Asia and Europe. In 1923 Piper and Morse reported 700 additional accessions, obtained mostly from China, Japan, Manchuria, and India. Since very few duplicates were received, the authors concluded that many local varieties existed in these countries. From 1924 to 1927 Dorsett collected 1500 types of soybean in Manchuria, and in 1929–1931 he and Morse collected 4578 varieties and types in Manchuria, Korea, and Japan. This was the only expedition ever made into eastern Asia specifically to collect soybean germ plasm.

By 1922 43 of the introductions received were found to be suited for seed production in the United States, and they were named. During the 20-year period 1907–1927, a number of introductions were received that were important as varieties or as parental material: AK, Biloxi, Dunfield, Haberlandt, Illini, Laredo, Mammoth Yellow, Manchu, Mukden, Peking, and Tokyo. Introductions from Nanking, China, in 1927 contributed directly, or after selection, the varieties Charlee, Clemson, Creole, CNS, Georgia, Missoy, Monetta, Palmetto, and Nanking. Although none of these varieties was grown extensively, Palmetto and CNS were valuable as parental material.

Approximately 10,500 introductions have been brought to the United States since 1898—many of these have been duplicates, so many have been lost. The current germ plasm collection maintained by the Plant Science Research Division of USDA, contains about 4,000 accessions including 200 named varieties. This collection represents a valuable source of genetic diversity if needed.

An important step in soybean variety development was the organization of the United States Regional Soybean Laboratory in 1936, with headquarters at the University of Illinois, Urbana, in cooperation with 24 north central and southeastern states. Informal cooperative research is also conducted with other states and Canada.

Since 1943 87 varieties have been registered by the Crop Science Society of America. This group includes varieties adapted throughout the soybean-growing area of the United States. For convenience, varieties are grouped into ten maturity groups (00 through VIII) corresponding to latitudinal bands. Group 00 varieties are adapted to and mature in the northern United States and southern Canada where the days are long. Group VIII varieties are adapted to the Gulf Coast area, where the days are short. The other maturity groups are distributed between these two extremes.

IMPORTANCE OF CROP

In 1970 42,447,000 acres produced 1,135,769,000 bu of soybeans in the United States. Ten states accounted for 85 percent of the 1970 production. The north central states of Illinois, Iowa, Indiana, Minnesota, and Ohio produced 58 percent of the soybeans harvested in 1970.

GENETIC UNIFORMITY

The vulnerability of any crop depends upon the uniformity of currently grown varieties. The practices that increase the vulnerability of varieties by restricting the germ plasm base are:

1. The parents used to develop new varieties are restricted to a few that tend to produce superior varieties under "normal" conditions.
2. A very few varieties dominate production.
3. Resistance to a given disease in currently grown varieties traces to a single source.

The currently grown soybean varieties have been evaluated by these three criteria.

To determine the diversity, the lineages of 62 currently grown varieties (43 northern and 19 southern) were traced. These 62 varieties have as ancestors the 29 introductions shown in Table 1. However, it is obvious that most of the currently grown varieties can be related to 11 introductions—Mandarin, Richland, AK, Manchu, Clemson, Tokyo, PI 54610, Mukden, Dunfield, Arksoy, and Roanoke. The sources of these 11 and their frequency of occurrence in currently grown varieties are given in Table 2.

Of the 62 varieties, only 17 were grown on 1 percent or more of the soybean acreage. In fact, the varieties Wayne, Amsoy, Corsoy, Clark or Clark 63, Lee, and Bragg accounted for 56 percent of the United States acreage. In the north central area (eight states), Wayne, Amsoy, Clark or Clark 63, and Corsoy account for 67 percent of the acreage. Three of these varieties have Richland as an ancestor, and all have Mandarin as an ancestor. The two southern varieties Lee and Bragg, which are grown on 58 percent of the Delta acreage, have both AK and Clemson as ancestors.

Another approach to evaluating genetic diversity is to look at the maternal ancestor alone. This permits consideration of the diversity of the cytoplasm as well as the nuclear component. This evaluation is particularly important in light of the corn blight situation. Evaluating essentially the same varieties as before, the frequencies of the introduced maternal ancestors are: in the North (eight states)—Mandarin, 74 percent; Illini, 18 percent; Mukden, 6 percent; Dunfield,

TABLE 1   Ancestors of 62 Currently Grown Soybean Varieties

Ancestor	No. Varieties Represented as an Ancestor		
	Northern^a	Southern^b	Total
Mandarin	36	0	36
Richland	26	0	26
AK	14	12	26
Manchu	24	0	24
Clemson (CNS)^c	4	13	17
Tokyo	6	11	17
PI 54610	6	9	15
Mukden	11	0	11
Dunfield	3	5	8
Arksoy	0	6	6
Roanoke	0	5	5
Kanro	4	0	4
Manitoba Brown	4	0	4
Jogun	3	0	3
Strain 171	3	0	3
Seneca	3	0	3
Flambeau	2	0	2
Peking	1	1	2
Haberlandt	0	2	2
Palmetto	0	2	2
Mammoth Yellow	0	2	2
Laredo	0	2	2
Korean	1	0	1
Swedish Strain	1	0	1
PI 65338	1	0	1
PI 70218-2-6-7	1	0	1
Otootan	0	1	1
Biloxi	0	1	1
PI 60406	0	1	1

^{a} Northern = maturity groups 00–IV.
^{b} Southern = maturity groups V–VIII.
^{c} CNS was selected from Clemson. Thought to be a mixture of Nanking and Clemson that was introduced the same time as Clemson (E. E. Hartwig, personal communication).

2 percent; in the South (six states)–Tokyo, 34 percent; Illini, 33 percent; Dunfield, 20 percent; Roanoke, 13 percent. Combined (85 percent United States acreage) this comes to Mandarin, 51 percent; Illini, 23 percent; Tokyo, 11 percent; Dunfield, 8 percent; Mukden, 4 percent, and Roanoke, 4 percent.

These statistics indicate that, for the soybean varieties currently grown, genetic uniformity is pronounced. The same germ plasm has been used repeatedly to develop varieties that produce well under normal conditions. Only when necessary has diverse germ plasm been

TABLE 2 Major Ancestral Varieties, Source, Year of Introduction into the United States, and Frequency of Occurrence in Parentage of Currently Grown Varieties of Soybeans

Ancestor	Source	Introduced	Frequency of Occurrence (percent)		
			Northern[a]	Southern[b]	Total
Mandarin	Pehtuanlintza, Northeastern China	1911	84	0	58
Richland	Changling, China	1926	60	0	42
AK	China	1912	32	63	42
Manchu	Niguta, China	1911	56	0	39
Tokyo	Yokahama, Japan	1901	14	58	27
Clemson	Nanking, China	1927	9	68	27
PI 54610	Changchun Liaoning, Prov. China	1921	14	47	24
Mukden	Mukden, China	1920	26	0	18
Dunfield	Fancheatun Stn., China	1913	7	26	13
Arksoy	Pingyang, Korea	1914	0	32	10
Roanoke (rogue from Pi 71597)	Nanking, China	1927	0	26	8

[a] Northern = Groups 00–IV.
[b] Southern = Groups V–VIII.

213

utilized. For example, Peking was the source of cyst nematode resistance and CNS the source of pustule resistance. Also, a number of the ancestors have been used in both the northern and southern varieties (Table 1).

Consequently, the three factors likely to increase the vulnerability of a crop are present in soybeans. The germ plasm has until now been sufficiently diverse to give stability to production, but the capacity to withstand attacks of new parasites in the future is uncertain. This is especially true where resistance to a parasite is traceable to one or two sources.

PESTS AND DISEASES

Fungi, bacteria, viruses, insects, and nematodes all attack soybeans and cause damage that reduces yield and quality. "Losses in Agriculture" (USDA, 1965) reports for the period 1951 to 1960 14 percent loss from diseases, 2 percent loss from nematodes, and 3 percent loss from insects. Losses caused by pests are often localized, and these relatively small areas may suffer extensively; other areas may be relatively free.

A complete list of the pests of soybeans is beyond the scope of this report. Only those pests of known or potential economic importance will be considered, and these are listed in Table 3.

STATUS OF BREEDING FOR RESISTANCE

Soybean vulnerability is directly related to the status of the breeding program. Regardless of the availability of resistance, the ultimate value is its incorporation into a high-production variety. Although the ancestors of current varieties are few, there has been sufficient diversity to give stability. Plant pathologists have recognized the potential of certain pests, and resistant varieties have been produced without economic loss to the grower. For example, target spot could cause as much as 50 percent loss, but because of resistant varieties the parasite does not reach its potential. Many genotypes from the germ plasm collection have failed to give material that yields as well as currently grown varieties, even when specific pathogens are troublesome. The repeated use of one or two introductions as the source of resistance has resulted in widespread distribution of the same germ plasm in currently grown varieties.

The major diseases are brown stem rot, pod and stem blight, purple seed stain, and virus diseases. Diseases such as *Phytophthora*

TABLE 3   Soybean Pests of Known or of Potential Economic Importance

Common Name	Scientific Name
Diseases	
Brown stem rot	*Cephalosporium gregatum*
Rhizoctonia rot	*Rhizoctonia solani*
Charcoal rot	*Macrophomina phaseoli*
Phytophthora rot	*Phytophthora megasperma* var. *sojae*
Target spot	*Corynespora cassiicola*
Downy mildew	*Peronospora manshurica*
Bacterial blight	*Pseudomonas glycinea*
Bacterial pustule	*Xanthomonas phaseoli* var. *sojensis*
Brown spot	*Septoria glycines*
Frogeye leafspot	*Cercospora sojina*
Purple seed stain	*Cercospora kikuchii*
Pod and stem blight	*Diaporthe phaseolorum* var. *sojae*
Soybean mosaic virus	Soybean mosaic virus
Bud blight	Tobacco ringspot virus
Nematodes	
Root-knot	*Meloidogyne* spp.
Soybean cyst	*Heterodera glycines*
Reniform	*Rotylenchulus reniformis*
Sting	*Belonolaimus gracilis*
	Belonolaimus longicaudatus
Insects	
Green clover worm	*Plathypena scabra*
Velvetbean caterpillar	*Anticarsia gemmatalia*
Soybean looper	*Pseudophisia includens*
Corn ear worm	*Heliothis zea*
Tarnished plant bug	*Lygus lineolaris*
Mexican bean beetle	*Epilachna varivestis*
Bean leaf beetle	*Cerotoma trifurcata*
Stink bugs (green)	*Acrosternum hilare*
Stink bugs (brown)	*Euschistus servas*
Seed corn maggot	*Hylemya platura*

rot and downy mildew have been controlled by resistant varieties. Other diseases, such as bacterial blight, are observed, but it has been difficult to show a significant loss from these parasites.

Lines with resistance to brown stem rot have been identified and are currently being evaluated for yield potential. Losses from brown stem rot occur throughout the soybean growing area; the severity of the loss is related to moisture conditions, since the organism impedes the movement of water within the host. There is also an interest in charcoal rot, which has symptoms similar to brown stem rot.

Phytophthora rot has been recognized as a serious threat to soybean production since about 1955. Two races of the organism are known. Available resistant varieties, of which Mukden and Arksoy are the sources, are adapted to all areas of the United States where the fungus is a threat.

The leaf diseases downy mildew, bacterial blight, brown spot, and frogeye leaf spot have not caused severe losses. Resistance is incorporated into new varieties where possible.

Pod and stem blight and purple seed stain are diseases most commonly associated with poor seed quality. Some of the current varieties have resistance, and breeding programs are under way to develop varieties with a higher level of resistance in areas where seed quality is a problem.

The viruses have not been investigated as extensively as have some of the other soybean parasites. Bud blight is sometimes very severe in localized areas. Soybean mosaic virus causes a seed coat mottling and is thought to be increasing in prevalence.

Most attention has been paid to the nematodes that burrow inside roots (root-knot and cyst-forming nematodes). Resistance to root-knot nematodes is available and has been incorporated into varieties for growing in the southern United States where this nematode is prevalent. However, more intensive studies have demonstrated that there are various races of this nematode.

The soybean cyst nematode was first discovered in the United States in 1954. All of the varieties grown at that time were susceptible. From a screening of the germ plasm collection, three resistant genotypes were found. One of these, Peking, was used to develop the resistant varieties that are now available. This resistance was satisfactory for three races of the nematode, but recently a fourth race has been reported. This new race severely damages varieties resistant to the other three. Currently the germ plasm is again being evaluated for resistance to Race 4. Preliminary observation indicates that resistance is available in the germ plasm pool.

Varieties resistant to the cyst nematode Races 1, 2, and 3 are also resistant to the reniform nematode. Limited screening is under way to develop resistance to several other nematodes that live outside the roots.

Numerous insects are associated with soybeans (Table 3). Resistance to some of these has been observed and breeding for resistance has begun.

The 4,000 accessions in the germ plasm collection provide a potential resource of resistance to those pests listed in Table 2 for which resistance is not now known. These sources need to be identified. Of equal importance, the collection needs to be augmented with more introductions, especially from mainland China. These additions would increase the diversity of the genetic resources available.

PEANUTS

ORIGIN

The peanut (*Arachis hypogae*) is a native of South America, possibly having originated in Bolivia. Archeological data from Peru indicate that the peanut has been under cultivation for about 3,500 years, but the plant was not known outside the Western Hemisphere until post-Columbian times. At the time of the discovery of America and European expansion into the New World, this cultivated species was known and grown widely throughout the tropical and subtropical areas of this hemisphere. The early Spanish and Portuguese explorers found the Indians cultivating the peanut in several of the West Indian islands, in Mexico, on the northeast and east coasts of Brazil, in all the warm land of the Rio de la Plata Basin (Argentina, Paraguay, Bolivia, extreme southwest Brazil), and extensively in Peru.

From these regions the peanut was disseminated to Europe, to both coasts of Africa, to Asia, and to the Pacific Islands. Eventually, it traveled to the southeastern United States, but the time and place of this introduction is unknown.

Apparently, the types of peanuts now grown commercially in the United States came to this country from Spain and Africa. Although peanuts have been grown commercially in the United States for at least 200 years, large-scale production first developed and increased during and after the Civil War.

IMPORTANCE OF CROP

Peanuts, grown on about 1,500,000 acres, rank ninth in acreage among the major United States crops. Production in 1970 was 3 billion pounds having a farm value of $375 million. About 54 percent of the acreage is of the Virginia and Runner types, 45 percent of the Spanish type, and less than 1 percent of the Valencia type.

Seven states (Georgia, Texas, Alabama, North Carolina, Oklahoma, Virginia, and Florida) account for 98 percent of the commercial peanut acreage. Within these states the major production areas are concentrated into a relatively few counties. Three-eighths of the United States production comes from the southern part of Georgia, and one-half from the contiguous area of south Georgia, southeast Alabama, and northwest Florida. An additional one-quarter is produced in the eastern counties of North Carolina and Virginia. Texas and Oklahoma grow slightly more than one-fifth of the United States crop, but Texas acreage is concentrated in two areas of the state and about half of Oklahoma's production is in one county.

GENETIC UNIFORMITY

Some 15 varieties are grown in the United States. Nine of these account for more than 95 percent of the peanut acreage, and of these, three varieties are grown on about 70 percent. One variety, Starr, comprises about 35 percent of the United States acreage. This was the first Spanish variety developed by selection after hybridization of two pure-line ancestors and, since its release in 1961, has rapidly displaced all of the pure line Spanish-type varieties previously grown. Now it is grown on almost all of the acreage in Texas and occupies two-thirds of the acreage in Oklahoma and nearly one-third in Georgia. This totals 77 percent of the area devoted to Spanish peanuts. Such a narrow base suggests marked vulnerability.

Most of the remaining Spanish peanut acreage is of the Argentine variety. Four new Spanish varieties released in 1969–1970 are under seed multiplication. Two of these were derived by simple line selection from Starr and Argentine. The other two have Argentine germ plasm as a common parent in their hybrid pedigrees.

All of these Spanish-type varieties are similar in adaptive range, growth habit, duration, water requirement, seed size, chemical composition, nutritional attributes, and end-use suitability.

Five related varieties of the Virginia botanical type (commercial Virginias and Runners) were grown on 644,000 acres (44 percent of the United States total) in 1970. This acreage comprises about 81.5 percent of the overall area for the type and is concentrated in the Southeast and the Virginia–Carolina belts. Florigiant and Early Runner are grown most widely of the five varieties. Florunner, released in 1969, has replaced a large area of its Early Runner parent during 1971.

The large-seeded Virginia segment of the Virginia botanical type has a restricted genetic base and could be very vulnerable to an epidemic of a major pest. About 79 percent of the Virginia market class trace to a common parentage from a cross made by Higgins in Georgia in the early 1930's between a small-seeded spreading peanut named Basse, collected in Gambia, and United States variety Spanish 18–38 (Table 4).

The narrow germ plasm structure of the peanut crop is evident from the predominant acreage of the Starr Spanish variety and varieties related to Florigiant Early Runner. These two populations include nearly 78 percent of the United States acreage. Moreover, the advanced breeding lines under evaluation as potential new varieties are heavily weighted with the Starr, Argentine, or Early Runner–Florigiant genotypes in their pedigrees.

There is very substantial evidence that the genetic base for peanuts has become increasingly narrower throughout the world wherever higher yielding varieties have replaced the numerous varietal strains once grown by farmers in local communities. This is cause for concern.

In 1970–1971 approximately 46 million acres of peanuts were sown in more than 51 counties (USDA, 1971). Farmers in India, China, the United States, Nigeria, Brazil, and Senegal grew 34.6, 15.3, 7.8, 4.8, 4.7, and 3.8 percent of the total world production, respectively. In India, with an average annual area of over 18 million acres, five states account for nearly 80 percent of the country's total production. A few varieties with more than a million acres each account for the bulk of India's crop. In Nigeria three primary varieties are in commercial production.

Four-fifths of Brazil's crop is concentrated in western São Paulo State, and 98 percent of this area grows the Tatu variety. Tatu is also grown in other states. More than 80 percent of the commercial and industrial activity of Senegal, where the crop constitutes the main source of income of the government, is directly concerned with peanuts. The 28–206 variety comprises almost half of the seed distributed by the agency responsible for collecting and preserving the seeds required for the annual sowing.

In other countries where marketing boards regulate the quality of export product or supply seed annually to growers, a single variety often predominates. About 99.5 percent of South Africa's million acres consists of the Natal Common peanut, a Spanish type. In Gambia, Malawi, and West Pakistan, single varieties account for more than half the commercial production.

Higher yielding peanut genotypes are often moved from one country to another, where they supersede the older varieties previously

TABLE 4 Narrow Germ Plasm Base in Cultivated Peanuts: Interrelationships of 13 Varieties Developed in Breeding Programs, 1930–1970

Year	Agency	Cross No.	Pedigree Data ($\female \times \male$)	Advanced Breeding Line No.	Variety Name	Year Released
1933?	Ga.	207	Basse × Spanish 18-38			
1933	Fl.	231	Fla. Sm. Wh. Span. 3x-1 × Dixie Giant	F231-51	Dixie Runner	1943
1933	Fl.	230	Fla. Sm. Sh. Span. 3x-2 × Dixie Giant	F230-118-9-5-1	Early Runner	1952
1944	Fl.	334A	Ga. 207-3 × F230-118-2-2	F334A-B	Florispan Run.	1953
1944	N.C.	B	Ga. 207-2 × Whites Runner	B-33	NC 2	1953
1947	Fl.	359	Jenkins Jumbo × F230-118-5-1			
1951	Fl.	392	F334A-5-5-1 × F359-1-3-14	F392-12-1-5-1	Florigiant	1961
1951	Fl.	393	F334A-3-5-5-1 × Jenkins Jumbo	F393-9-5-1-2-2,4	NC 17	1969
1951	Ok.	None	Line selection from Pi 121070	OK P-2	Argentine	1951
1955	Ga., USDA	None	Line selection from PI 121070-3	GA T-1271; OK P-112	Spanhoma	1969
1958	Ga., USDA	CI	Argentine × Spanish 18-38-42	CI-27	Tifspan	1970
1958	Ga., USDA	C32	Argentine × *Archis monticola*	C32S	Spancross	1970
1959	Ga., USDA	None	Line selection from Sp. 18-38	Sp. 18-38-42	Spanette	1959
1960	Fl.	439	F334A-3-14 × F230-118-B-8-1	F439-16-10	Florunner	1969
1962	Ga., USDA Israel	C133	Florigiant × F334A-B-17-1	Israel X-30	Shulamith (Israel)	1968

220

grown. The Shulamith variety, which since its release in 1968 has substantially replaced the former varieties cultivated in Israel, is an Israeli selection from a United States cross having Florigiant as one parent. The United States variety Starr has been recommended to replace its Spantex parent in Australia.

Despite the hundreds of genotypes in peanut germ plasm collections in each of the major producing countries, fewer than a score of varieties comprise the bulk of the world's production. Several factors contribute directly to the trend toward fewer varieties and increased vulnerability through the hazard of genetic uniformity. The peanut is 94–100 percent self-pollinated (Hammons, 1971). The usual breeding techniques for self-pollinators lead to morphological uniformity, since all varieties must be uniform, distinct, and stable. Seed multiplication fields are inspected and certified as uniformly true to type, with near-zero tolerance for off-type variants in foundation, registered, and certified seed generations.

Growers prefer to produce the one variety that returns the highest yield at the lowest cost per unit of production. In the United States, grading standards for the minimum price support schedule penalize the grower of mixed peanuts. Warehousemen and shellers prefer to handle one or at best a few varieties, since separation of types and quality segregations add to their costs. Other segments of the industry directly or indirectly encourage a narrow genetic profile by seeking to standardize quality attributes of end-use products. Processors have traditionally used blends consisting of varying proportions of the Virginia and Spanish types to obtain products, particularly peanut butter, with desired characteristics of stability, texture, and flavor.

PESTS AND DISEASES

From the worldwide view, leafspots (caused by *Cercospora arachidicola* and *C. personata*) are the most serious diseases. An in-soil pod rot or pod breakdown, caused by *Pythium* and *Rhizoctonia* spp. and possibly other soil-borne fungi, results in severe economic losses in certain parts of the United States and is increasingly prevalent and damaging in other countries. A seedling and later season blight, caused by *Aspergillus niger*, is a serious problem in many of the important peanut-producing countries of the world. In certain areas rosette virus and bacterial wilt, caused by *Pseudomonas solanacearum*, sharply limit production. Peanut rust, caused by *Puccinia arachidis*, is a serious disease in the West Indies, where it is endemic, but has not proved to be the threat once envisioned. However, in recent years rust has appeared with increasing frequency and destructiveness in south-central Texas.

A large number of insect pests find the growing plant, the maturing fruits, or the stored seed an attractive source of food. Species of several genera of plant parasitic nematodes cause root, peg, and pod injury.

Average United States annual losses caused by diseases have been estimated at 28 percent of production, with destruction from insects and nematodes adding another 6 percent (USDA, 1965). Breeding resistant genotypes has been a powerful means of pest control in other major crops. However, none of the present commercial varieties has appreciable resistance to the principal diseases, insect pests, and nematodes that attack the crop. Moreover, little useful genetic resistance to such pests has been found in the cultivated peanut species. Certain wild Arachis species are immune or highly resistant to several major diseases or pests including the *Cercospora* leaf spots, rosette virus, peanut stunt virus, rust, northern root knot nematode, and a mite. The wild species with such characteristics are in sections of the genus *Arachis* whose species cross rarely if at all with cultivated peanuts. Therefore, the use of such resistance requires as yet unknown methods of interspecific hybridization.

There are opportunities to improve insect and disease control through breeding in *A. hypogaea,* but information is lacking not only on good sources of resistance but also on the genetic basis for the resistance that is found. Present evidence suggests that insects show nonpreference for certain peanut genotypes under experimental conditions (Leuck *et al.,* 1967), but the genetic nature of such nonpreference feeding is unknown.

A physical basis for leafspot resistance—smaller stomata—has been reported for two varieties without any evidence concerning its inheritance, and little apparent progress has resulted from efforts to incorporate the resistance into higher yielding varieties. Recent research has revealed physiological resistance to peanut rust in three different peanut introductions, but no information is available on inheritance of this reaction.

The Schwarz variety is a classic example of the use of resistant germ plasm to save the peanut industry from destruction by a bacterial wilt caused by a virulent strain of *Pseudomonas solanacearum.* Wherever bacterial wilt is severe, cultivation of Schwarz 21, or the Matjan variety derived from it, is the most satisfactory means of control. Inoculation trials with other isolates of *P. solanacearum* indicate relatively low susceptibility to bacterial wilt in at least one United States commercial variety. The genetic basis of resistance is unknown.

Genetically determined resistance of cultivated peanuts to rosette virus disease has been conclusively demonstrated, and resistance has been incorporated by backcrossing to develop the new Senegal variety 28–206–R.R. This is the only instance in peanut research where genetic information has been available to achieve effective control of a major disease by breeding.

The above examples by their meager number sharply define the critical shortage of sound genetic data by which the peanut breeder could meet a threat to the industry posed by a new strain of parasite. The absence of good resistant varieties has forced the peanut industry to rely on pesticides.

STATUS OF BREEDING

United States peanut breeders are revising old procedures and devising new techniques to broaden the genetic base of the crop. Many of the recently released varieties were derived from selection following hybridization between varieties in the Virginia and Spanish subspecific groups of *A. hypogaea* (Table 4). Changes in early generation selection procedures have given varieties that are composed of a number of lines that breed true for certain characters.

The peanut is a tetraploid species that has undergone chromosome doubling during its evolutionary history. Many qualitative characteristics, especially in crosses between subspecies, are complexly inherited, with interacting systems of duplicate genes (Hammons, 1971). Such genetic systems promote a higher degree of heterogeneity among sibs in advanced breeding lines.

Peanut geneticists are screening a varied collection of genotypes, many of which were collected as primitive land races in the South American countries near the center of origin for the genus *Arachis.* Where high levels of tolerance to diseases, insects, or nematodes have been found, breeding is in progress to incorporate such resistance into higher yielding adapted varieties.

This diversification should result in populations with greater resistance to epidemic hazards than their purebred ancestors.

DRY BEANS

ORIGIN, HISTORY, AND IMPORTANCE

On the basis of diversity, and on the evidence of discovery of the indigenous wild forms, the region of origin of the common bean, *Phaseolus vulgaris,* can be identified as the southern portion of Mexico, Guatemala, Honduras, and part of Costa Rica. The oldest remains of beans associated with human cultural debris are found in caves in the Tehuacan area of central Mexico, carbon-dated at about 5000 B.C., and in pre-Incan tombs of Peru dating to about 3000 B.C.

Beans, along with maize and squash, comprised the staple diet of the ancient peoples of upland Central America, from which the plant spread to North and South America. The common bean and the lima bean, *Phaseolus lunatus,* were widely distributed in the Americas and the Caribbean in pre-Columbian times.

Dry bean production in the United States amounts to some 18 million 100-lb bags annually, grown on about 1.5 million acres in 14 different states. The farm value of the crop in the United States in each of the last two years, 1969 and 1970, has amounted to an average of $150 million. Almost 90 percent of the acreage and an equivalent portion of the total United States production is accounted for by six states; Michigan alone produces about 35 percent of all dry beans in the country on 43 percent of the total acreage in beans.

PRODUCTION PATTERNS AND EPIDEMIC POTENTIAL

Several features of the production pattern are relevant to the potential for epidemics of diseases or insects in the United States bean crop. In the first place, the bean production areas in this country are widely dispersed from New York to California, but in the major producing states the bean acreage is distributed continuously in a single region of each state. Furthermore, among the major producing areas there is only a minor overlap in the varieties being grown. For example, New York produces Light Red Kidneys and the Black Turtle bean; Michigan grows chiefly the navy bean, with Dark and Light Red Kidneys a distant second; Nebraska and Colorado grow mostly Pintos and Great Northerns, whereas California divides its acreage among lima beans, blackeyes (cowpeas), the small white, and the Dark and Light Red Kidney beans.

Two commercial types, the Michigan navy bean and the Pinto, account for 60 percent of all dry beans grown in the United States. Two Pinto varieties, U.I. 111 (University of Idaho 111) and U.I. 114, account for almost all the acreage of this type. Four navy beans, Sanilac, Seaway, Gratiot, and Seafarer, account for nearly all the acreage of this class. Pintos U.I. 111 and 114 are closely related, the 111 being one parent of 114. All four navy beans, despite complex breeding pedigrees, derive over 90 percent of their germ plasm from the previously widely grown navy variety Michelite, whose resistance to common bean mosaic virus (bean virus 1) came originally from the old Michigan Robust variety.

It is clear, therefore, that, for a considerable part of the edible dry bean acreage in the United States, annual production rests upon a dangerously small germ plasm base. This is particularly true for the navy bean. The situation is less critical for the Pinto class because the production is dispersed in several different states, whereas the navy beans are produced in a single concentrated region—the Saginaw Valley and adjacent peninsular or "thumb" region of Michigan.

More than half of the states which account for nearly 40 percent of the acreage and 50 percent of the country's production are situated in the western United States, where the summer rainfall pattern is semiarid to arid, the humidities are low, irrigation is necessary, and the virus diseases are generally more important than are bacterial blights. For the other states, situated in the prairies (excepting Nebraska) and the Great Lakes Basin, summer rainfall is depended upon for production, humidities are higher than in the West, and the bacterial blights are more important than the mosaic diseases, though the latter are not excluded.

Since several of the major bean diseases are seed-borne: common blight, caused by *Xanthomonas phaseoli;* fuscous blight, caused by *X. phaseoli* var. *fuscans;* halo bacterial blight, caused by *Pseudomonas phaseolicola;* anthracnose, caused by *Colletotrichum lindemuthianum;* and the BV1 and BV15 strain of the common bean mosaic virus. Thus a seed production industry to guarantee disease-free seed is essential until disease-resistant varieties can be bred and brought into production. At the present time, the picture is encouraging. Resistance to curly top virus, mosaic caused by bean viruses 1 and 15, halo blight, common bacterial blight and bean anthracnose is available in several released commercial varieties.

All dry beans, with the exception of Tara and Jules, are susceptible to the common and fuscous blights. All colored dry beans, except

Redkote, are susceptible to halo blight. The Pinto and navy beans, though slightly susceptible upon artificial inoculation, possess a high degree of field resistance to halo blight.

The nationwide dispersal of the various bean-growing regions provides a safeguarding isolation of one region from another such that an epidemic arising in one region is not likely to extend to another. The major qualification to this assessment involves seed transmission of disease organisms or insects from a seed-producing region (Idaho or California) to a seed-purchasing region.

A second level of protection serving to retard the spread of disease from one region to another is provided by the different spectra of varieties grown in the various regions. To be sure, some varieties are common to two or more regions (Table 5), but the extent of commonality is diminishing as each region produces replacement varieties more specifically suited to the agronomic or disease problems of the region. For example, the Royal Red Kidney solves the virus problem for eastern Washington, but not the halo blight problem of Michigan or New York. New York and Michigan, on the other hand, will soon solve their halo blight problem with the release of a line or lines resistant to this disease. In this manner, genetic divergence in the Red Kidney group will come about with respect to both major genes and polygenes controlling nonspecific resistance.

The greatest potential for a regional disease or insect epidemic lies within each region, because it is here that the varieties of a given commercial class are most closely related, the acreage concentration is greatest, and the environment promoting a particular parasite is most homogeneous. The most conspicuous example is in Michigan, where over 500,000 acres are planted each year to four closely related navy bean varieties derived from the older variety Michelite. Though these varieties all carry resistance to the prevalent races of anthracnose and common bean mosaic, they are uniformly highly susceptible to common blight, including its variant form, fuscous blight. And although it cannot be wholly confirmed, it is probable that the serious regional outbreaks of common and fuscous blight in Michigan in 1969 and 1970 can be attributed in part to the varietal homogeneity. Contributing factors have been the shorter rotation schedules, approaching monoculture in many cases, the persistence of the bacteria in the soil and plant debris, and the favorable temperature and humidity coincident with the period of greatest susceptibility. Unfortunately, there is not nearly enough western-grown or certified seed to plant the entire Michigan acreage. Consequently, disease-contaminated seed is

TABLE 5 1970 Harvested Acres, Showing Extent of Acreages of Given Commercial Classes in Common (figures are in thousands)

Producing States

Commercial Classes	Cal.	Col.	Ida.	Kan.	Mich.	Minn.	Mont.	Neb.	N.M.	N.Y.	N.D.	Ut.	Wash.	Wyo.
Navy					505									
Red Kidney	56		1.3		41					51.6			1.7	
Great Northern			15.5			2.2	0.9	60.2						5.2
Pinto		240	52.3	23	20.5	13.1	10.1	25.8	4.0		24.5	17.0	8.5	24.8
Pink			21.8			2.5					0.5		3.4	
Small red			13.1										17.0	
Black turtle										25.8				

227

often used, making localized disease outbreaks almost a certainty and a widespread epidemic more likely.

GENETIC DIVERSITY

Since the common dry bean, the snap beans, and lima beans reproduce by self-pollination, every variety grown in the United States is a pure line or an assemblage of closely related pure lines—all plants in the variety are genetically identical or highly related. Every field of dry beans of a given variety, therefore, is genetically and cytoplasmically uniform or nearly so, depending upon the exact procedure followed by the breeder during variety development.

Nevertheless, there is enormous genetic diversity within species of the common and the lima bean. Practically all of the diversity lies between varieties. There must be, of course, additional variation *within* varieties that are mixtures of types—such as primitive varieties or "land-races" indigenous to the regions of origin. Only a small fraction, therefore, of the total possible diversity in the species is actually found among the varieties in prevailing use today in the United States. The requirements of dry and snap bean processors for beans of a given size, seed color, or canning quality and the preferences of growers for beans of a particular plant form and maturity range have been met by the breeder by breeding to a type; that is, by narrowing the range of variability so that each successive variety released bears a close similarity to the preceding standard. To the extent that breeding to a type results in genetic uniformity for factors affecting disease or insect susceptibility, the variety and the region throughout which the variety becomes widely grown is exposed to possible disaster should a new virulent race of a fungus, bacterium, strain of a virus, or insect pest arise or be introduced.

The Pinto bean closely rivals the navy bean in total production. The major varieties are the University of Idaho lines 111 and 114, which are closely related, and approximately half of the total acreage is in Colorado. It might be reasonably inferred that the Pinto bean crop is potentially vulnerable, but additional factors enter the picture. For one, the total Pinto acreage in the United States is distributed in 12 of the 13 western states that grow beans. Second, these states are in the subhumid to semiarid zones of the country where the diseases that are dependent upon high humidity, rain, and hail for dissemination are not generally prevalent. In general, also, the incidence of the fungus and bacterial seed-transmitted diseases in these regions is exceedingly low. The major production hazards are the mosaic and curly top viruses, and insect pests. Protective levels of genetic resistance to the viruses are becoming available, and as yet, the destructive insects are controllable by pesticides. Where moisture, precipitation patterns, and temperature conditions favor a particular disease, however (e.g., halo blight in Idaho in the middle 1960's), a serious epidemic is possible even in the West. Sprinkler irrigation, rain showers, and hail are conducive to such local outbreaks in the western states.

Genetic uniformity is conducive to epidemics, but genetic diversity is no guarantee of protection. For example, alfalfa is highly heterogeneous genetically, and yet the spotted aphid was able to generate severe epidemics throughout the West until specific resistance was found. The European corn borer in maize, and the cereal leaf beetle of small grains provide similar examples of the failure of genetic diversity to protect major crops from damage. What is required of crop genetic systems is that there be diversity of the genes concerned with host resistance, both for the genes governing direct interactions with the parasite, and for those genes governing a more indirect or general resistance. Fortunately, the presently available genetic resistance to common bacterial blight of beans is polygenic, and once the polygenes are brought to sufficiently high frequency, may confer adequate resistance. By exploiting polygenic sources of resistance, breeders are not required to seek out and incorporate specific genetic resistance factors against specific races of the parasite. The specific factors suffer the fatal disadvantage that the resistance they confer is sometimes short lived, breaking down in time because of replacement of present races by other virulent races.

STRATEGY OF VARIETAL DEVELOPMENT

Consider, as an example, the question of building a line of defense against common bacterial blight. At present the only genetic defense consists of a set of polygenes that, individually, have but small effect on the trait under consideration; it traces to a "rogue" plant in the Great Northern Nebraska 1 variety. These polygenes will be incorporated into the standard acceptable types of a region by a combination of backcrossing and intercrossing and the products released for commercial production. The varieties Tara and Jules represent this first line of defense for Great Northern beans in Nebraska. But alternative and independent sources of polygenes conferring resistance should be sought, perhaps in plant introductions, and if found, introduced into the same standard types. The resulting varieties would comprise the second line of defense. However, they should not be held in abeyance until the first line gives way to variant forms of the bacterium. Rather, they should be ordered into battle at once on two possible fronts, representing two strategies of genetic diversity. Under one strategy the new lines would be introduced as varieties, equivalent to the earlier varieties in resistance, type, agronomic worth, etc., to be grown alternately on different fields throughout a region, or alternately by the same grower on a given field in a rotational sequence with a variety of the first group. A complementary strategy would require blending seed of the two kinds into a common lot to be grown as a composite variety.

If the two sets of genes that confer resistance are independent of each other, it should be possible to incorporate both sets into a single variety. This has been done in certain varieties for resistance to anthracnose.

For the normally self-pollinating species, these suggestions represent all that can be accomplished in building genetic diversity into crop varieties. The use of hybrid bulk populations or complex composites, even if they are appropriately diverse for disease or pest-resistance factors, may well introduce unacceptable diversity for agronomic and quality factors.

GENETIC CONTROL OF RESISTANCE TO SPECIFIC DISEASES

Common Mosaic In the case of common mosaic, genetic resistance to BV1 and BV15 strains of the virus has been incorporated into most dry beans widely grown in this country. Two sources of mosaic resistance are used in the United States. The first, discovered in the Robust navy bean in 1915, is recessive and is effective against BV1. It has been widely used in dry bean types and has afforded protection for over 50 years, engendering no epidemic of another disease. The other source was Corbett Refugee bean, which carries a single dominant gene effective against BV1 and its variant BV15. This gene has been known since 1929 and has been transferred to many other varieties. There are strains of the common mosaic virus to which neither the recessive Robust gene nor the dominant Corbett Refugee gene confer resistance. Fortunately, these viruses have not yet been reported to occur in the United States. If they were introduced they could cause severe epidemics.

Anthracnose In the United States, three races—alpha, beta, and gamma—of the anthracnose organism must be taken into account. (A fourth race, delta, has been reported in the United States and found virulent on over 50 commercial United States and European varieties of snap and dry beans. Genetic resistance to delta was found in two noncommercial lines. The prevalence of delta in the United States is not known). Fortunately, genetic resistance has been discovered for each of the first three races, and the modes of inheritance have been established. The genetic bases of resistance varies from simple to complex, depending on the varieties involved and the particular race of the organism under consideration.

The simplest inheritance pattern is found for Race alpha; there is one gene, the dominant form of which confers resistance. But some crosses reveal the presence of two dominant genes, on different chromosome segments, either one of which makes for resistance. For Races beta and gamma, the inheritance pattern becomes more involved, in that additional genes are required and their modes of action more complex.

It appears that, in these patterns, as many as 10 different sites (loci) in the chromosomes—can govern resistance or susceptibility of the plant to the three races. The accumulation of favorable genes at all 10 loci would be a formidable task, and in fact no known variety possesses this favorable combination. Certain P.I. strains, however, circumvent the problem very simply through possessing the single gene *Are,* which confers resistance to all the races simultaneously. Whether this gene can be incorporated without difficulty into commercial types, and how long the resistance might last are unknown at present.

Halo Bacterial Blight It is premature to conclude that the general pattern of inheritance of resistance to halo blight is well understood. For one thing, the question of races of the bacterium remains clouded. With particular halo blight strains, evidence can be obtained from a given segregating population favoring a single major gene, with resistance due to the recessive condition. By contrast, other crosses suggest that resistance is affected by modifiers. Additional research discloses resistance due to a single major dominant gene. It has also been shown, using different material under field nursery conditions, that several genes operate to determine a range of resistance levels. Quantitative inheritance has also been indicated in particular populations. Thus, the only general principle that can be stated for halo blight resistance is that there appear to be protective levels of genetic resistance in several different varieties and that the frequency with which the relevant genes are expressed is high enough to permit the breeding of resistant varieties in any class or type of bean desired. Perhaps, when the apparent inconsistencies of inheritance have been clarified, it will be feasible to incorporate into otherwise similar varieties alternative genetic systems conferring resistance to halo blight. It might even be possible, as has been done for anthracnose resistance in certain varieties, to build into a single variety at least two alternative systems for resistance, so that the organism would have the much more difficult genetic task of simultaneously overcoming a double line of defense in the host.

Common Bacterial Blight   A useful level of genetic resistance to common blight has been found in the common bean in only one source, certain late-maturing, off-type plants of the Great Northern Nebraska 1. Genetic studies conducted by Coyne and Schuster of Nebraska indicate that, in crosses with G.N. 1140, these rogue plants contribute several genes of minor effect to the hybrid. In advanced self-pollinated or backcross generations, the segregational pattern suggests polygenic inheritance for resistance, with heritability estimates in the range of 54 percent to 67 percent. We urgently need additional and more effective sources of resistance to common blight, and in particular to the fuscous blight which already constitutes about 50 percent or more of the isolates studied. One source of resistance to fuscous blight is in the tepary bean, *P. acutifolius,* but the species barrier to gene transfer to *P. vulgaris* is formidable, having been breached only once in thousands of crosses.

Bacterial Wilt   While not now widespread, bacterial wilt, caused by *Corynebacterium flaccumfaciens,* does occasionally result in serious losses. No resistant commercial dry bean is known. Tolerance in some Great Northern 1 individual plant selections was shown by Coyne *et al.* (1965) to depend upon two complementary genes resulting in tolerance.

Root Rot   Beans have long been grown in the presence of *Fusarium* root rot caused by *Fusarium solani* f. *phaseoli* without suffering a major regional epidemic. Nevertheless, from New York to California root rot poses an ever-present threat. No commercial varieties are resistant, though some are claimed to be slightly tolerant. Genetic resistance has been found in a few tropical and Brazilian black beans, and in the Scarlet Runner bean, *Phaseolus coccineus.* In the tropical black bean, P.I. 203958 for example, the resistance is attributable to some two or three major recessive genes, plus numerous minor genes. In certain crosses, both a recessive gene pair and a dominant gene, distinct from the first pair, are necessary for resistance. In crosses involving the runner bean source, there is evidence of a system of dominant genes conditioning resistance.

These different patterns of resistance raise the possibility, discussed under "Strategy of Varietal Development," of incorporating both genetic systems for resistance, the recessive and the dominant type, in a single variety or in similar varieties to be used interchangeably in the same area.

Rust Rust can cause widespread destruction and is found in both the humid and dryland regions of the United States. It exists as a large number of physiologic races, and varieties of beans show resistance to some races but not others. The genetic basis of resistance has not been established on any comprehensive scale. For a few races, one dominant gene in the host determines resistance; for other races, more than a single gene was required for resistance. There is simply not enough known at present to speculate on the possible major genes for broad resistance to many races, or of polygenic resistance that would be difficult for the parasite to overcome.

Nonparasitic Diseases It should be noted that diseases and insects, while clearly of predominant importance, are not the only hazards to which genetic uniformity may expose a particular crop. The uniform response of a variety to mineral deficiency or toxicity can seriously impair performance and grower acceptance of a variety, or require special management by the producer in order to overcome the deficiency. Zinc deficiency had been noted in several of the bean-producing regions of the West, and extensively in Michigan. Of special concern in Michigan was the discovery that Sanilac, the leading navy bean variety in the state at the time, displayed necrotic leaf lesions, retarded growth, and poor yields throughout some 200,000 acres in the heart of the bean area. This prompted the release of the variety Saginaw, which had proven capable of normal growth and near-normal production on these sites. At about the same time it was learned that the cause of failure of Sanilac was zinc deficiency, a problem best solved by adding zinc to bean fertilizer. Varietal diversity could not have overcome the problem permanently, but earlier and wider use of the tolerant Saginaw in the zinc-deficient area could have reduced the disastrous yield losses while the real cause was being sought.

PRESERVING A GENE OR A POLYGENIC SYSTEM

How best can individual, perhaps rare, genes or polygenic complexes, with unknown potential value, be preserved until needed by a breeder to solve some production or quality problem? Bean seed is not particularly long-lived even under the most ideal storage conditions.

It has been proposed that, for economic reasons, bulk populations be established—germ plasm "pools" into which would be placed representative samples of each of a large number of introductions, varieties, elite lines, etc. Different pools would be formed to serve various needs or geographic areas, and they would, indeed, be relatively simple and inexpensive to maintain. The critical question, however, is whether such pools would, in fact, maintain the rare gene or favorable group of genes where interplant competition prevails. Differential fecundity from both genetic and environmental causes, including sampling, makes it virtually certain that some genes will be lost—either completely from the pool, or beyond easy retrieval because of infrequent occurrence.

While maintenance may be simple and inexpensive, ultimate retrieval of a desired gene or polygenic system from a pool may prove costly both in resources and time. Many breeders would prefer that identity of the individual line be maintained wherever possible, even if this means that the breeders themselves have to share the task of maintenance by periodically growing a portion of the collection and returning small seed lots to the Fort Collins, Colorado, facility for storage.

The foregoing situation applies to both the dry and snap beans; for a comprehensive discussion of the latter, see the following section.

SNAP BEANS

HISTORY AND IMPORTANCE

In his catalog published in 1822 Thornburn listed eight varieties of garden beans, *Phaseolus vulgaris.* Shortly after 1900 some 185 distinct varieties were grown in the United States. The earliest record of

FIGURE 1   Snap bean picker harvesting 38-in. rows in Wisconsin. (Photo courtesy W. C. Gabelman, Univ. of Wisconsin.)

a round-podded snap bean was about 1865, when German Black Wax was introduced from Germany. It was also the first wax-podded bush variety. Now nearly all of our varieties are round podded.

In 1870 Keeney of LeRoy, New York, selected Stringless Refugee Wax, the first stringless bean. Prior to that time, all beans were stringy, more or less fibrous, and of only fair-to-medium quality. The majority of snap bean varieties introduced since then in the United States have been stringless and many of these possibly involve the above variety in their pedigree. No stringy variety would be accepted by the food trade today and only varieties whose pods are practically fiberless are used for processing.

In 1970 86,000 acres of snap beans were harvested for the fresh market and 228,000 for processing; their farm value came to over $95 million. Most of the seed of the bush bean type is produced in irrigated districts of southern Idaho while the seed of the pole bean type comes from California under the control and supervision of seedsmen. Snap bean seed production is virtually limited to districts of those two states because of the freedom from the seed-borne diseases, anthracnose, and several bacterial blights. The industry was moved from New York, its first center, in the mid-1800's, because anthracnose and common bacterial blight caused heavy losses in the East and Middle West where rain and humid weather favored their development and spread. Resistance to anthracnose and the blights has not been developed in any commercial snap bean variety, but the diseases are at present of only minor importance in the United States.

About 35.5 million pounds of seed of the green-podded bush types, 4.5 million pounds of wax-podded bush types, and approximately 2.7 million pounds of pole beans were produced in 1970. Most of the pole beans, with the exception of the Blue Lake pole variety, which is processed, are used for fresh market and in home gardens. Bush-type beans are used primarily for processing and the fresh market.

GENETIC UNIFORMITY

Like dry beans, every snap bean variety is a pure line. This makes all plants within a variety essentially uniform genetically. The available diversity, then, is between varieties.

Seedsmen now produce seed of more than 70 varieties of snap beans; however, only a few types, including fewer than 20 varieties, are grown in substantial quantities for processing and the fresh market. Of the 35.5 million pounds of seed of green-podded bush types produced in 1970, about 16 million pounds (46 percent), consisting of about 12 varieties have Tendercrop germ plasm in their ancestry; about 5 million pounds (15 percent) of about 8 varieties have Blue Lake germ plasm, and another 5 million pounds (15 percent) of 5 varieties have Harvester germ plasm. Thus, about 76 percent of the total seed produced of all green-podded bush varieties possess germ plasm from only three major sources. The varieties within a class resemble one another closely. Varieties with Tendercrop and Blue Lake germ plasm are used almost exclusively for processing, and those with Harvester germ plasm are most popular as fresh market varieties.

Thus, the principal diversity in snap beans lies among these three groups. More diversity would be desirable.

The wax-podded bush varieties have more genetic diversity. Of a total of 4.5 million pounds of seed produced of 5 varieties, slightly over 1 million pounds (24 percent) possess Wisconsin Wax 536 germ plasm. One variety in this class comprises about 20 percent of the total seed produced.

It is evident that much of the bean crop grown for processing and fresh market may be potentially very vulnerable to an epidemic of a new disease or insect pest because of the narrow genetic base of the varieties grown. Consider that 72 percent of the processed crop is grown in the East and central states and that 53 percent is centered in three states—Oregon, Wisconsin, and New York—the latter two having a similar environment during the growing season. Also, 40 percent of the fresh market crop is produced in Florida and an additional 34 percent in several of the other southern states. All this makes the crop particularly vulnerable.

Since the snap bean seed-growing industry is centered in an area environmentally different from that of most of the processing and fresh-market growing areas, each could be subject to different disease or insect epidemics. If a disease epidemic became widespread in southern Idaho, where most of the seed of the bush-type varieties is produced, it could seriously jeopardize the national supply of snap beans for all purposes. Just such a situation threatened that area from 1963 to 1967, when epidemics of halo bacterial blight occurred in 1963 and 1964; for a time it appeared the district might lose the enviable status as a blight-free seed-producing area that it had enjoyed for over 50 years. Unusual weather occurred in 1963 and 1964: more than average rainfall, several hail storms, and relatively low temperatures during the growing seasons. All this favored the development and spread of the halo blight organism. In 1963 the disease was observed in about 10 percent of the total acreage. The seriously infected fields were plowed under, and where the infection was less severe, diseased plants were removed or infected areas burned. Stringent eradication measures were put into effect in 1964 and since then, little if any halo blight has been observed. However, this instance shows the danger of relying on a single seed-producing area for the production of most of the seed used for the entire country.

MARKET REQUIREMENTS

Buyers of processed and fresh-market beans are very specific in their demands concerning the type and quality of beans they will accept. This is a major reason why a narrow genetic base underlies most snap bean varieties now grown.

An acceptable processing variety must have white seeds; round, dark green, fiberless pods of medium length; firm flesh; medium-early maturity; mosaic resistance; and high yield. A fresh-market bush-type bean must have white seeds, round and medium-length pods with a slight amount of fiber necessary to maintain quality in shipping, and an upright-growing bush. This insistence upon a narrow range of uniformity of characters causes genetic vulnerability to pest attack. Furthermore, since practically all bush beans for processing are mechanically picked, the number of types is further restricted, for such varieties have to be adapted for machine harvesting. Plants must be sturdy and upright, with the pods well dispersed throughout the plant and of fairly even maturity. This additional genetic uniformity adds to the vulnerability of the variety.

Mutations, a source of genetic diversity, have played an important part in the early development of snap beans and were responsible for bush habit, round and wax pods, and stringlessness (Zaumeyer, 1963). Some of our most popular present-day varieties arose from mutants, whose plant and pod characters were essentially the same as those of the present varieties, although their seed color is the cream or white desired by the trade. Gallatin 50, for example, one of the most popular of the Tendercrop types, is a mutant selected from Tendercrop.

DISEASES AND INSECT PESTS

The principal diseases of snap beans (Zaumeyer and Thomas, 1957) and the sources of resistance to many of them are given in Table 6. The average annual loss caused by diseases is approximately $20 million. Although several insects, including the Mexican bean beetle (*Epilachna varivestis*), spider mites (*Tetranychus telarius*), and thrips (*Thrips tabaci*) can damage beans, no variety shows any resistance to these pests. They are kept under control with pesticides.

None of the American snap bean varieties used commercially before 1920 was disease resistant. However, bean anthracnose is of no particular concern in snap beans, and the bacterial blights are of minor significance in most years, since essentially disease-free seed

TABLE 6 Principal Bean Diseases and Resistant or Tolerant Germ Plasm

Disease	Cause	Resistance (R) or Tolerance (T)	Source
Common bean mosaic And strains	Virus	R	Corbett Refugee and many commercial varieties
Bean yellow mosaic	Virus	Specific R	Great Northern 31 and Scarlett Runner
Curly top	Virus	R	Burtner's Blightless and many other varieties
Common blight	*Xanthomonas phaseoli*	T	Great Northern Nebraska 1, Jules, and Tara
Fuscous blight	*X. phaseoli* var. *fuscans*	R	Tepary bean
Halo blight	*Pseudomonas phaseolicola*	R	Redkote, Pinto U.I. 111 and 114, Great Northern U.I. 16 and 31
Brown spot	*P. syringae*	—	—
Powdery mildew	*Erysiphe polygoni*	Specific R	Stringless Green Refugee, U.S. 5 Refugee, Topcrop, and others
Root rot	*Fusarium solani* f. *phaseoli*	T	P.I. 203958, P.I. 165426, and P.I. 165435
Root rot	*Pythium butleri*	R	P.I. 203958
Root rot	*Rhizoctonia solani* f. *phaseoli*	—	—
Root rot	*Thielaviopsis basicola*	R	P.I. 203958, NY 2114-12
Rust	*Uromyces phaseoli typica*	Specific R	Tenderwhite, Custer, Seminole, Pinto U.S. 5
Southern blight	*Sclerotium rolfsii*	—	—
White mold	*Sclerotinia sclerotiorum*	—	—
Root knot	*Meloidogyne hapla*	R	P.I. 165426, P.I. 165435

239

is used by practically all growers. Some breeding work to develop resistance to halo blight is being conducted.

Common Mosaic Virus Most of our presently grown snap bean varieties resist the common bean mosaic virus, including the New York 15 strain, and all varieties used for processing and fresh market are resistant to both viruses. With the exception of the Blue Lake varieties, their mosaic resistance traces to one variety, Corbett Refugee, a selection made in 1929 from mosaic-susceptible Stringless Green Refugee.

Since resistance originated from only one source, its genetic vulnerability may appear somewhat dangerous, although it has not broken down in over 40 years. Similar resistance to the New York 15 strain has been maintained for the same period. Corbett Refugee confers resistance to all strains of the common bean mosaic virus prevalent in the United States.

Curly Top Resistance to the curly top virus has recently been incorporated into five snap bean varieties. These varieties also possess resistance to the common bean mosaic virus and to the New York 15 strain. Curly top resistance, which has been shown to be dominant and conditioned by two factors, was originally isolated from a noncommercial bean known as Burtner's Blightless. A number of strains of the virus have been discovered that do not infect the resistant varieties.

Curly top is transmitted by the beet leafhopper (*Circulifer tenellus*), the only known vector of the disease, and is present only in certain western arid sections of the United States where this leafhopper is found. The Columbia Basin of central Washington is such an area. The disease is so widespread there that only a variety immune to the disease can survive. In other respects it is an ideal growing area for the production of bean seed free of the bacterial blights and anthracnose. Since southern Idaho now produces about 80 percent of the bush snap bean seed in the United States and may at some time be vulnerable to a disease or insect epidemic, it would be wise to develop another equally good bean-producing area. The introduction of curly top and mosaic-resistant bean varieties will allow a new seed-growing area to be developed elsewhere, thus alleviating the hazard of growing too large a portion of the seed required in a single area.

Yellow Mosaic Virus Many strains of bean yellow mosaic virus are found wherever beans are grown. No snap bean varieties resist most strains of the virus; the Scarlett Runner bean, *Phaseolus coccineus,* is resistant to more strains than any variety of common bean. The inheritance of resistance from that variety was reported to be governed by a single dominant factor; others reported it to be inherited in a recessive manner conditioned by two or three major genes, with additional modifiers affecting symptom expression.

Resistance to the severe pod-distorting strains of bean yellow mosaic virus in Great Northern U.I. 31 was found to be conditioned by three major recessive genes, with additional modifiers.

Aphid resistance in the Black Turtle Soup variety was reported to be related to field resistance to the virus. This type of resistance should be effective to all strains of the virus carried by the vector.

Halo Bacterial Blight Idaho halo blight epidemics of 1963 to 1967 stimulated considerable research on bacterial diseases of bean in the United States, particularly on halo blight itself. No commercially acceptable snap bean variety resists this disease, although many of the dry bean sorts show field resistance.

Root Rots Tolerance to *Fusarium* root rot is now being incorporated into snap bean varieties, although the task is proving to be very difficult. No thoroughly effective chemical, biological, or genetic control has yet been developed.

Breeding beans for root rot resistance has been hampered by failure to find a high level of resistance in bean germ plasm. A degree of tolerance to *F. solani* f. *phaseoli* is available in P.I. 203958, a native bean collected in Mexico, and in P.I. 165426 and P.I. 165435, also from Mexico. The last two have also been reported to be resistant to *Rhizoctonia solani* and to the root knot nematode (*Meloidogyne hapla*). Resistance to *Fusarium* root rot superior to that found in the common bean is also found in *Phaseolus coccineus,* but no commercial varieties have been released with root rot resistance derived from any of the above sources. Resistance to *Pythium* blight, caused by five species of *Pythium,* was recently found in P.I. 203958.

Rust Bean rust, caused by *Uromyces phaseoli typica,* has been reported from all parts of the world. In the past it has been of importance on dry beans in some of the mountain states, but currently it is of significance on snap beans in fall-grown crops along the Atlantic Seaboard.

No varieties have been found to resist all of the 34 reported races of rust. The most resistant snap bean variety is Tenderwhite; others showing considerable resistance are Custer, Earligreen, Florigreen, Olympia, and Seminole.

Studies on the inheritance of resistance to six races of the rust organism showed that resistance to Races 1 and 2 is governed by a single factor, but that more than one factor is involved in resistance to Races 6, 11, 12, and 17.

SOURCES OF GERM PLASM

The National Seed Storage Laboratory at Fort Collins, Colorado, maintains stocks of many commercial bean varieties as well as many older genetic stocks. The USDA Plant Introduction Station at Pullman, Washington, maintains a world bean collection of over 8,000 items as well as several thousand accessions of related *Phaseolus* species, some of which can be hybridized with *P. vulgaris.* The USDA Regional Pulse Improvement Project in Tehran, Iran, also maintains a large collection of foreign bean varieties.

The Bean Improvement Cooperative (BIC) is an organization of bean investigators throughout the world. Through the BIC annual reports a bean breeder can obtain breeding stocks and related information from many bean breeders and pathologists throughout the world.

A sample of all released bean varieties and other useful germ plasm, such as segregating material furnished by plant breeders and foreign varieties, should be preserved in a permanent depository as the National Seed Storage Laboratory to insure their preservation for future generations.

MEASURES TO REDUCE VULNERABILITY

Although high-quality snap beans for canning, quick freezing, and the fresh market are essential, it appears that the varietal specifications set by the buyers of beans grown for these uses may be too rigid, thus unnecessarily narrowing the desirable range of genetic variability. Why only white-seeded types producing slender, dark green, round pods of a particular flavor are acceptable is difficult to determine. Although colored-seeded varieties produce a slightly discolored liquor when canned, if they are processed after seed color develops, they can be used for quick-frozen products of a very high quality. Pods of colored-seeded varieties, when slightly over-mature, can be distinguished from those of frozen white-seeded sorts only by slightly discolored seed coats and there is no impairment of quality.

Before the introduction of the Harvester-type beans, most market varieties were colored-seeded. High quality of seed in the latter varieties is generally more readily maintained than in white-seeded sorts, and germination of colored-seeded varieties under cool, wet soil conditions is usually better than that of white-seeded ones. If the specifications for acceptable snap bean varieties were less demanding in such unimportant characters as pod roundness and seed color as long as high eating quality and yield were retained, the genetic diversity among varieties could be greater and hence they could be less vulnerable to a disease or insect epidemic.

The fact that 80 percent of the country's supply of bush snap bean seed is produced in a relatively small area of southern Idaho is extremely dangerous and creates a potentially very vulnerable situation for the users of this seed throughout the United States. Since the Columbia Basin of central Washington is also an ideal bacterial blight- and anthracnose-free seed-growing area, but is subject to serious curly top infection, greater effort should be made to breed additional curly top-resistant varieties for seed production in this area. This would allow more of the southern Idaho acreage to be diverted to the Columbia Basin, which has thousands of acres of excellent land and thus would help ensure the bean growers of the country a dependable seed supply. When additional acreage for processing is needed, the Basin would also be an ideal area for expansion.

Because of the low humus content of the soils in the Columbia Basin, *Fusarium* root rot develops rapidly even on virgin soils if good crop rotation is not consistently practiced. It has been demonstrated that growing beans two or three successive years on the same land can produce almost 100 percent root rot infection. This same disease also causes considerable loss to beans (6 percent) in other bean-growing areas of the United States if growers do not adhere rigidly to crop rotation. Efforts should be made to incorporate, as rapidly as possible, resistance to *Fusarium* root rot into as many snap bean varieties as possible.

Since many dry bean varieties resist the halo blight organism, breeders should make more effort to develop resistant snap bean varieties. Such varieties would be of particular value if this disease again becomes of importance in southern Idaho.

Bean rust can be serious in winter-grown beans in Florida and the fall-grown crop along parts of the Atlantic Seaboard and adjacent areas. Resistance to many of the rust races is available, and development of acceptable varieties resistant to as many races as possible should be started.

Bean breeders should attempt to develop the nonspecific type of resistance to diseases, or specific resistance based on at least two or more genetic factors. The likelihood of a double mutation in a parasite to overcome uniform resistance is less than the probability of overcoming a single factor in specific resistance.

A continuous search for natural mutations should also be made as another source of genetic diversity.

PEAS

HISTORY

The origin of peas (*Pisum sativum*) as a human food is obscure. Garden pea domestication is so ancient that its wild prototype has never been found. Either the wild type no longer exists, or, through evolution, the modern pea has become very unlike its original ancestor. A major difficulty in tracing the history of garden peas is that until recent times no distinction was made between green or garden peas and field peas or, for that matter, between several legume species such as peas, beans, vetches, lentils, chickpeas, and lupines that were often referred to simply as "pulses." Many references to peas in old writings of the Greeks and Romans indicate that peas were widely used in the Old World. Explorers of Stone Age Swiss lake villages found evidence of pea use by Stone Age and Bronze Age man. No doubt peas were used as human food in prehistoric times. The pea was an early food crop in Britain and it was introduced early to the American continent. Peas were sown in 1602 on the Elizabeth Islands off the coast of Massachusetts.

No history of peas, however brief, would be complete without mention of the work of Gregor Mendel (1822–1884). Through experiments with peas, he laid the foundation for most of the modern work on inheritance and may be said to have established genetics as a science. Mendel was a monk in the Franciscan monastery at Brünn, Austria. In the cloister gardens there he made crosses between varieties of peas that differed in plant height, flower color, seed color, and other traits. By keen observation and maintenance of accurate pedigree records he demonstrated that inheritance of characters is not a hit-or-miss affair; it follows certain definite patterns. His results were published in an obscure journal in 1866, but the paper was overlooked or ignored until 1900 when three scientists (Correns, de Vries, and Tschermak) independently "discovered" Mendel's paper and proclaimed its importance in the field of plant and animal breeding.

Most of the early pea breeding of record was done by English plant breeders who did much to improve plant type and quality. Pea breeding did not become important in the United States until this century, particularly after 1920.

Early seed production in the United States was concentrated in Michigan and New York State. Prevalence of seed-borne foliage and pod diseases resulted in a gradual shift, beginning about 1910, to the semiarid and arid irrigated regions of the West, where it is possible to produce seed much freer of most seed-borne diseases, particularly *Ascochyta* blight (*Ascochyta pisi*), *Mycosphaerella* blight (*Mycosphaerella pinodes*), and bacterial blight (*Pseudomonas pisi*). Use of pea seed grown in dry western areas has resulted in a reduction of these diseases so that they are currently of minor importance in the United States.

Pea seed production is currently centered in irrigated areas of Idaho, Washington, California, and Montana, and the natural rainfall region of eastern Washington and northwestern Palouse area of Idaho.

IMPORTANCE OF CROP

In the United States peas are grown for use as a fresh vegetable, for canning and freezing, for drying, and, to some extent, for forage and cover crops. Most seed is for crops used for processing as either canned or frozen peas. Small amounts of seed are used for home garden, market garden, and fresh shipping. In 1970 about 384,000 acres of peas were harvested green, including 252,000 for canning and about 132,000 for freezing.

Dry edible peas are a good source of protein, and production of these types is greater than production of seed peas used for the fresh market or processing. Three hundred ninety-five million pounds of dry edible peas were produced in 1970.

Seed of Austrian Winter peas has been grown in the United States primarily for forage and cover until recent years when the dry peas have been shipped to Japan and other countries to be ground into flour and used as a protein-rich food supplement. The flour is dark and is not readily accepted as food in the United States. Most of the Austrian Winter peas are grown in Idaho and in 1970 about 95 million pounds of seed were produced.

The value of the pea crop to the farmer in 1970 totaled approximately $80 million.

GENETIC DIVERSITY

Since the advent of the green pea viner (sheller) and large-scale processing of green peas, first by canning and later also by freezing, the types of peas developed by breeders have, by necessity, fallen within rather narrow limits.

Peas for processing fall into a few general and rather all-inclusive classes. These are the smooth-seeded or "Early June" types made up almost exclusively of the variety Alaska and derivatives of it; Surprise and Early Sweet types, which are wrinkled-seeded but, in some cases, possess Alaska germ plasm; and "main season" types that are true "sweets" or wrinkled-seeded types.

The canner and freezer have been able to market peas with only certain specific qualities; the canning types must have light green and the freezing types dark green "berry" or green pea color. Recently a few canners have been processing peas with dark green "berries." Although color is different for the two styles of processing, the plant type suitable for the two processes is similar. In addition, both types react similarly to pea diseases. For yield and cultural reasons, ease of shelling, etc., acceptable types have been limited to those with determinate vines, double pods being preferable to single pods. A few varieties have more than two pods per fruiting node.

The breeding and selection of peas for use by processors is dictated by four primary requirements: Does the variety have acceptable quality? Is its yield satisfactory? Is is adapted to commercial methods of handling and processing? Last but not least, does it possess resistance to the pests and diseases it may encounter?

In 1914 and 1915 a pea line selected from the variety Advancer and named Davis Perfection was introduced to the canning trade. It proved to have such an acceptable vine and podding habit with good processed (canned) quality and good yielding ability that it was very popular and eventually became the standard by which other processing varieties were judged. This variety (and later disease-resistant strains of it) was widely used by pea breeders in developing varieties for both canning and freezing purposes, and there is now a series of canning varieties of varying season and a similar series of varieties with freezer pea color that have Perfection germ plasm.

A recent survey of commercial pea breeders, workers in state experiment stations, and USDA personnel included the question whether they could name any main season variety in use by processors in the United States at the present time that *did not* contain Perfection germ plasm. None was forthcoming.

The major varieties of early peas used by processors are Alaska or types with Alaska germ plasm.

The consensus of pea breeders is that, from the 1970 pea seed crop, approximately 60 million pounds contain Perfection germ plasm, and 37 million pounds have Alaska germ plasm, for a total of 97 million pounds. Thus 96 percent of the total of the green pea seed produced for use by processors have one of but two types of germ plasm.

Of the 395 million pounds of dry edible peas produced in 1970, about 257 million pounds were of Alaska types (green-seeded) and much of the balance was of the "First and Best" variety (yellow-seeded). These peas are marketed either as whole or split peas.

The genetic base for peas appears to be narrow. Few pea breeders are developing new varieties for processor use, and many varieties are selections from an established type or the result of renaming a variety to suit the whims of the seed merchant.

Peas are self-pollinated, so all plants of a given variety tend to be genetically uniform. While this is desirable in the production of a useful economic crop, it also means that all plants of a variety are equally susceptible to unforeseen hazards.

Within the common pea species, there is enormous genetic diversity but its use is restricted because of the limiting factors of trade acceptance. The commonly used varieties represent but a small part of the total known genetic variation available in the species.

Since one of two germ plasms is present in most pea varieties used for processing today, one can assume a considerable degree of genetic vulnerability to disease or insect attack. Indeed, the disease outbreaks in peas during the past 40 years bear out this assumption.

DISEASES AND INSECT PESTS

The average annual loss caused by diseases to green peas is slightly over $12 million and to dry edible peas about $2.75 million. Major diseases of peas are listed in Table 7. Plant breeders have incorporated resistance, or are in the process of doing so, with regard to several pea diseases, in order to provide commercially acceptable resistant types.

The average annual loss caused to green peas by insects is slightly over $2 million, to dry edible peas about $1 million. The primary insect pests of peas are several species of aphids, several species of foliage-feeding caterpillars ("loopers"), the pea weevil (*Bruchus pisorum*), and the pea moth (*Grapholitha nigricana*). The most troublesome of the aphid species is *Macrosiphum pisi*, but other species are

TABLE 7 Pea Diseases and Availability of Resistant or Tolerant Germ Plasm

Disease	Cause	Resistance (R) or Tolerance (T)	Source
Wilt	*Fusarium oxysporum* f. *pisi* Race 1	R	Many varieties
Near wilt	*F. oxysporum* f. *pisi* Race 2.	R	Delsiche Commando and many other varieties
New wilt	*Fusarium oxysporum* f. *pisi* Race 5	R	P.I. lines
Root rot	*Fusarium solani* f. *pisi*	T	P.I. and USDA lines
Root rot	*Pythium ultimum*	T	P.I. and USDA lines
Root rot	*Rhizoctonia solani* f. *pisi*	—	—
Root rot	*Aphanomyces euteiches*	—	—
Ascochyta blight	*Ascochyta pisi*	T	P.I. lines
Mycosphaerella blight	*Mycosphaerella pinodes*	T	P.I. lines
Soft rot	*Sclerotinia sclerotiorum*	—	—
Bacterial blight	*Pseudomonas pisi*	R	Stratagem
Powdery mildew	*Erysiphe polygoni*	R	Stratagem
Downy mildew	*Peronospora vicae*	T	Few varieties
Anthracnose	*Colletotrichum pisi*	—	—
Common pea mosaic	PV 2	R	Many varieties
Enation mosaic virus	PENM	R	Types developed from P.I. 140295
Seed-borne mosaic	PSBMV	R	P.I. 193586 and P.I. 193835
Streak inducing virus complexes		T	P.I. lines

248

also instrumental in transmitting virus diseases. The loopers cause damage by foliage feeding and are especially troublesome since they are difficult to remove from shelled peas and may occasionally be found in the finished food product, whether canned or frozen.

Some varietal differences in rate of aphid build-up have been reported, and one variety of dry edible peas (Blue Prussian) is reported to be less susceptible to weevil infestation. However, no commercially available varieties possess appreciable resistance to any of these insect pests.

GENETIC VULNERABILITY

Since a great deal of breeding has been done to incorporate resistance to diseases into peas and since the source of resistance has often been from germ plasm that can be termed "exotic," it is apparent that, even though Perfection or Alaska germ plasm is present, it may be rather dilute in presently used varieties. Perfection or Alaska germ plasm should not be construed as being detrimental, *per se*, since over the years these types have time and again proven their dependability. The point at issue is the matter of narrowness of genetic foundation in relation to hazardous exposures of all kinds. Thus, the genetic base of peas may not be as narrow as first appears. It is doubtful if peas, as a crop, are as vulnerable as corn appears to be, but this may be because no exotic pea parasite has made its appearance.

There is a tendency on the part of some pea breeders to look upon the arbitrary classes of peas, e.g., Early Sweets, Perfections, Alaskas, as units of inheritance unto themselves. These distinctions are practical and useful, and they provide a convenient form of reference; but these types are not in fact inherited as packages. Each has certain common characteristics, but underlying these apparent similarities there may be countless subtle differences. It is possible to use genetically diverse parent materials and, after breeding and selection, to end up with a type that fits nicely into one of the familiar so-called types. From external appearances, such breeding lines so resemble one of the established types that one would be led to believe they are genetically similar; that is, have a narrow genetic base. Yet, these lines may differ from the established type by many genes. The end product, the variety, is an assemblage of countless genes that may be arranged in an infinite number of configurations.

Some pea varieties bred in the regions of the United States where downy mildew and certain leaf roll diseases are not a serious threat were found to be tolerant to resistant to these diseases when planted in Europe and Australia. There was no conscious effort on the part of the pea breeder in the United States to produce resistance to these diseases, which is evidence of the hidden variation that may not be immediately apparent in pea varieties.

Spontaneous mutations occur within existing commercial pea varieties. Roguing, that is, destroying the off types, will reduce the number of obvious mutations; but many of the less obvious may persist in the variety, thus adding to genetic diversity.

With respect to the narrowness of cytoplasm source, the corn breeders were dependent on a single source of cytoplasm in order to achieve the desired goal of male sterility. There is no evidence that such a condition exists in peas. Quite the contrary, pea breeders often make reciprocal crosses, thus introducing more than one type of cytoplasm into their new lines.

The conclusion must be drawn that, in the case of green peas, whether for seed or for processing, the germ plasm base is narrow enough to present some danger in the case of attack by a disease new to the United States pea-growing areas or a new and virulent race or strain of an existing disease. However, there is valuable diversity as a result of the addition of disease-resistance factors from widely separated germ plasm sources. The evidence suggests that there is probably more genetic diversity in existing varieties of peas used in the United States than might be supposed if too much emphasis is placed on the percentage of Perfection and Alaska germ plasm apparent from variety production data.

In the case of dry edible peas, the genetic base is as narrow as each of the few varieties used for this purpose, since little breeding has been done in these types.

Several factors combine to lessen the danger of such disastrous crop losses in peas as occurred in the 1970 corn crop. One of the most important factors is the geographic distribution of pea acreages. Generally, the areas used for production of pea seed are distinct from areas used to produce crops for processing. There are a few western areas where both seed and processing acreages intermingle, but other areas produce either seed or raw produce for canning and freezing, but not both.

Pea-processing areas are scattered, primarily in the northern tier of states, from Maine and the eastern shore of Maryland to the Skagit Valley of northwestern Washington.

Pea seed acreages are somewhat more concentrated in the irrigated valleys of the intermountain regon, and dry edible peas are largely grown in the Palouse region of eastern Washington and northwestern Idaho.

Another factor is related to the low-"fold" increase of peas compared to many other crops. This can be both an advantage and a hazard when considering genetic vulnerability. The average-fold increase of peas when grown for seed is approximately 10 times. For processing acreages, the rate of seeding ranges from 200 to 300 lb of seed per acre, a relatively large amount of seed. Acreages for seed production must be figured at roughly 10 percent of the anticipated processing acreages. It is obvious that any appreciable loss experienced in the seed crop can have a drastic effect on the next season's acreage of crops for processing, unless adequate reserve stores of seed are maintained—and there are economic reasons that preclude this as a regular practice.

Low-fold increase can be an additional obstacle when it becomes necessary or desirable to release a new variety as soon as possible. Seed multiplication and varietal release can be speeded by producing "winter" or off-season increases in other regions, and this is a common practice among pea breeders and seed producers.

The advantage of low-fold increases in seed production is that the plant breeder developing new pea varieties is limited in the speed with which he can progress, and thus he has more time to assess his developments and become aware of faults and weaknesses of a new line before it is released. Obviously, the breeder can check his new lines only for known hazards.

An encouraging fact is that, within the past few years, pea breeders in the United States have become more aware and interested in pea varieties bred in England, Holland, Germany, and elsewhere, with the result that germ plasm of these types is being added to the genetic diversity of United States varieties.

MEASURES TO REDUCE VULNERABILITY

Looking to the future we must ask: How can the germ plasm base for peas be broadened in order to lessen the danger of disastrous epidemics? What steps should be taken to assure availability of useful germ plasm for all crops?

In view of the risks in having too narrow a germ plasm base, breeders should be encouraged to breed as much diversity into pea lines as is compatible with good agronomic characteristics and quality requirements.

Interested persons should try to assure continued research into all phases of pea breeding, pea diseases, pea production, etc., by state, federal, and private agencies.

The work of the National Seed Storage Laboratory should be supported. The activities of the Plant Introduction Service of the New Crops Research Branch, USDA, should be encouraged and supported. In this connection, the regional coordinators of this program should not be required to provide continuing and dramatic evidence of how these collections are being utilized as a prerequisite of their survival. They should be charged with the responsibility of maintaining this valuable reservoir of germ plasm, carefully documenting origin, and having it readily available to all breeders in the event of need.

The Regional Plant Introduction Station at the New York State Agricultural Experiment Station, Geneva, is providing a repository for all pea germ plasm. This collection is broad and diverse, having been assembled from all available sources and includes all germ plasm whether it appears to be of practical significance or not. This effort should be supported and encouraged along with the work of the Pisum Genetics Association, which is attempting to catalog all pea genes and make this information available through publication of "The Pisum Newsletter."

REFERENCES

Coyne, D. P., M. L. Schuster, and J. O. Young. 1965. A genetic study of bacterial wilt (*Corynebacterium flaccumfaciens* var. *surantiacum*) tolerance in *Phaseolus vulgaris* crosses and the development of tolerance to two bacterial diseases in beans. Am. Soc. Hortic. Sci. Proc. 87:279–285.

Garren, K. H. 1959. The stem rot of peanuts and its control. Va. Agric. Exp. Stn. Tech. Bull. 144. 29 p.

Hammons, R. O. 1971. Inheritance of inflorescences in main leaf axils in *Arachis hypogaea* L. Crop Sci. 11:570–571.

Leuck, D. B., R. O. Hammons, L. W. Morgan, and J. E. Harvey. 1967. Insect preference for peanut varieties. J. Econ. Entomol. 60:1546–1549.

Piper, C. V., and W. J. Morse. 1910. The soybean: history, varieties and field studies. U.S. Dep. Agric. Bur. Plant Ind. Bull. 197.

U. S. Department of Agriculture. Agricultural Research Service. 1965. Losses in agriculture. Agric. handb. 291. 120 p.

U.S. Department of Agriculture. Foreign Agricultural Service. 1971. World agricultural production and trade, statistical report, May 1971. p. 15–19.

Zaumeyer, W. J. 1963. Some new Tendercrop mutants. Seed world 92:6.

Zaumeyer, W. J., and H. R. Thomas. 1957. A monographic study of bean diseases and methods for their control. U.S. Dep. Agric. Tech. Bull. 868. Revised ed.

CHAPTER 14

Vegetable Crops

Contents

NATURE OF VEGETABLE CROPS	254
FACTORS CONTRIBUTING TO GENETIC	
VULNERABILITY	257
Uniformity	257
Resistance to Parasites	259
Nutritional Quality	259
Marketing Conditions	260
GENETIC VULNERABILITY BY CROPS	261
The Cabbage Family	261
The Nightshade Family	262
The Vine Crops	264
Leafy Vegetables	265
Onions and Carrots	266
SUMMARY	267

253

Vegetables are represented by many species and include annuals, biennials, and perennials. The commercial product—a wide variety of such edible plant parts as roots, leaves, petioles, fruit, seed, and floral parts—is generally low in dry matter, and therefore unique among major crops. Vegetables are not usually eaten as a source of energy but rather for their vitamin and mineral content as well as salad value. Many are eaten fresh after cleaning or peeling; others are cooked. But in all cases they are consumed directly by man. The current value of these crops to the farmer exceeds $2 billion annually in the U.S. (see Table 1).

NATURE OF VEGETABLE CROPS

Vegetable production started primarily as a garden culture in which man grew only enough for his own needs. A few species could be stored fresh, but most required processing. With the advent of commercial pasteurization, canning, and freezing, and a highly sophisticated distribution system, specialized industries developed in geographic areas where product quality and yield, product cost, and distribution efficiency could be optimized. These were usually in tem-

TABLE 1   Value of Vegetable Crops (farm level) in the United States, 1970

Crop	Fresh ($1,000)	Processed ($1,000)	Total ($1,000)
Cabbage	77,273	4,700	81,973
Broccoli	28,973	N.A.	28,973
Cauliflower	23,045	–	23,045
Brussels sprouts	7,147	N.A.	7,147
Tomatoes	205,941	170,653	376,594
Peppers	44,651	–	44,651
Eggplant	3,936	–	3,936
Cucumbers	30,499	55,443	85,942
Cantaloupes	82,760	–	82,760
Honeydew melons	10,925	–	10,925
Watermelons	56,303	–	56,303
Sweet potatoes	55,021	(inc. in fresh)	55,021
Carrots	66,635	(inc. in fresh)	66,635
Celery	85,620	–	85,620
Lettuce	223,276	–	223,276
Onions	101,010	(inc. in fresh)	101,010
Spinach	7,024	6,510	13,534
Escarole	7,850	–	7,850
Sweet corn	68,268	44,736	112,004
Asparagus	21,242	34,203	55,445
Red beets	–	4,412	4,412
Peas	–	51,889	51,889
Snap beans	40,505	54,363	94,868
Lima beans	–	14,198	14,198
Potatoes (1969)	–	–	622,000
Total	1,247,904	441,107	2,311,011

Source: USDA Stat. Rep. Serv., 1970 Annual Crop Summary – Vegetables for Fresh Market.
USDA Stat. Rep. Serv., 1970 Annual Crop Summary – Vegetables for Processing.

perate climates near market centers. The southern states met the demand for winter production; the arid West (with irrigation) was right for many crops in which high dry matter is important, e.g., onions to be dehydrated, or where arid conditions tend to decrease foliage diseases, e.g., muskmelons. The Pacific Northwest, the Great Lakes states, and New York were appropriate for cool-season crops marketed or processed during the summer and fall months. As in other crops, intensified culture brought many problems, particularly those concerned with parasites.

Vegetable production started as a hand operation. Cultivation was mechanized through use of equipment and row widths (36–38 in.) dictated by the size of the omnipresent horse. Planting and spraying operations were mechanized subsequently on the same traditional 36-in. row. Elimination of the hand weeding that supplemented me-

FIGURE 1 Broccoli in California. (Photo courtesy Asgrow Seed Co., Milford, Conn.)

chanical tillage awaited the successful development of effective herbicides. At first, mechanical harvesters were of little value for most vegetables but now varieties have been developed with uniform characteristics and precise periods of maturation. They are a boon to the grower and reduce the price of the commodity to the consumer, but their very uniformity makes them vulnerable to parasites and weather.

New systems of vegetable culture have evolved. Industry now adjusts rapidly to incorporate novel and useful systems of agriculture. Rows have been reduced to 9–12 in. apart, but these high-density plantings are increasingly vulnerable to the unknowns of the environment. Breeders find it difficult to change varieties quickly enough to neutralize the new problems resulting from changes in technology.

Vegetable crops are high in water content; therefore, they provide a medium that is attractive to many parasites—insect, fungus, bacterium, and nematode. In areas where monocultures of a single crop, such as watermelons, cabbage, or onions, developed, the problem of parasites of all types rapidly became acute. Chemical control was not adequate. It is not by accident that Orton was the first to demonstrate the potential of disease control through breeding resistant varieties in watermelons and cotton and that Jones and Walker extended this principle to many crops, using a model derived from studies on the yellows disease (caused by *Fusarium conglutinans*) in cabbage. Their work on biological control of disease predated Rachel Carson's *Silent Spring* by a half-century. This technique has been so successful that now a prime objective of every major vegetable breeding program is resistance to known parasites.

Resistance, however, is often not complete. In cabbage yellows, resistance has been adequate and stable; but in diseases like club root of cabbage, caused by *Plasmodiophora brassicae*, resistance is restricted to certain races of the organism, and there are many races. In other diseases like *Fusarium* wilt caused by *Fusarium oxysporum* f. *lycopersici* in tomato, resistance was adequate until a second race of the organism appeared. The intensive culture of resistant varieties tends to provide a selection pressure that favors mutants of the parasitic organism. Vegetable breeders have used genetic controls to minimize the effects of diseases. The growers have tempered their production practices to maximize genetic control and minimize chemical treatment, always recognizing the need for adequate controls and often the need for both systems.

FACTORS CONTRIBUTING TO GENETIC VULNERABILITY

UNIFORMITY

An overriding requirement of a modern vegetable variety is uniformity, especially of time of harvest. Successful culture requires uniform germination and subsequent growth, and precise cultural practices; but the degree of uniformity is limited by the degree of genetic homogeneity of the population of plants that constitutes the variety. Seeds must give rise to plants with equivalent capacity to respond to their environment with similar rates of growth. Plants that lag behind by as much as a day or two often act as "weeds" to the main crop. Homogeneity is valuable only so long as all traits of the variety are good and the variety totally adapted to the area of culture. But if any interaction between the vegetable and its environment is undesirable, the variety is uniformly undesirable. Thus, homogeneity (or uniformity) is a recognized two-edged sword. Yet the competitive nature of industry permits no alternative.

Variety homogeneity results from:

1. Reproduction from a single plant by vegetative cuttings, e.g., potatoes.

2. Reproduction by seeds that result from continuous self-pollination, e.g., lettuce and tomatoes.

3. Reproduction by seed resulting from controlled cross pollination (first generation hybrids), e.g., hybrid onions and hybrid cucumbers.

Of the 25 vegetables listed in Table 1, only sweet potatoes and Irish potatoes are propagated by cuttings. These two crops are discussed in Chapter 12.

A special type of genetic vulnerability exists in the self-pollinated crops. Industry has tended first to accept and later to demand certain varieties; a relatively few varieties of lettuce and peas have become predominant. For reasons not always apparent, new varieties of lettuce must be "Great Lakes" types and primarily composed of the Great Lakes variety germ plasm. Similar examples can be cited for Perfection-type peas, Tendercrop beans (Chapter 13), and California Wonder peppers. The process of continuous self-pollination eliminates nearly all genetic variability in a given variety. If old varieties are not maintained, or if old varieties are not sufficiently different from the type varieties of commerce, the only sources of new germ plasm will be the wild, exotic species. The development of new varieties from exotic germ plasm requires many generations of hybridization and selection. Thus, the change toward monocultures of varieties tends to eliminate or severely restrict the amount of reserve genetic material available either for commercial use or in a form that approaches commercial quality for breeding in the self-pollinated crops. Since the evolved varieties are both highly inbred and highly homogeneous, they are particularly susceptible to mutant parasites, the ubiquitous change of technology, and seemingly ever more stringent regulatory requirements. If change in varieties necessitates the use of exotic germ plasm, the change will require either time (in terms of plant generations) or adequate, continuous backstop programs in plant breeding by public agencies to ensure reasonable reserves of new variety material. The latter does not exist today.

Hybrids of cross-pollinating species are also vulnerable because of their uniformity. However, if a reasonable array of parent inbreds exists, industry can shift to new hybrids rather quickly.

The systems by which hybrid seed is produced commercially provide a source of vulnerability that can be serious. The potential instability of inbred cabbages, dependent on self-incompatibility for large-scale commercial hybridization, is already well documented. The dependence on male-sterile cytoplasm in onion, carrot, table beet, and sweet corn for commercial hybrid seed production parallels the dependence on T cytoplasm in field corn. In fact, the model for hybrid seed production using male-sterile cytoplasm was first developed in onions by Jones and Clarke.

Normal cucumber plants produce both male and female flowers, but there are two genes that in combination produce gynoecious cucumbers (all flowers female). The cucumber industry has shifted to hybrid seed production, using these gynoecious inbreds as seed parents, a shift that has taken place within the last decade. Only time will tell if industry was justified in making this wholesale change involving two genes. Since the sex of cucumbers is so readily altered by two growth-regulating chemicals, ethephon and gibberellin, it follows that parasites of cucumbers might also react to these basic, natural biochemical differences and become more serious than before.

RESISTANCE TO PARASITES

Nearly every important variety of vegetable carries resistance to one or more diseases. In addition, some carry resistance to insects and nematodes. Resistance is a highly desirable attribute, but should be used in conjunction with other control methods, including chemicals. Parasites can mutate to new forms that, if more virulent, soon become dominant. At that point a search for new genes for resistance must be made. If the source of resistance is from an exotic species, the problems of attaining acceptable quality in the derived varieties will be difficult. The degree to which industry has become dependent on genetic resistance is documented by specific examples later in this discussion.

NUTRITIONAL QUALITY

Breeders of crops consumed directly by humans have been greatly concerned about the nutritional quality of their product. Unfortunately, facilities for assay have been woefully inadequate. If the breeder's wife likes the new variety, or if the children will eat it, or if limited taste panels by industry representatives approve, the variety is often deemed suitable for release and public consumption. Many of these same breeders recognize the possibilities of antimetabolites* that only animal feeding tests or tedious chemical methods will detect. They are aware, for example, of the USDA request to have potato variety Lenape removed from public use because of its excessive alkaloid content. Breeders are also becoming aware of the Food and Drug Administration's revision of Part 121, 121.3, which requires affirmation as generally recognized as safe (GRAS) by the Commissioner, "Substances that have had significant alteration of composition by breeding or selection and the change may reasonably be expected to alter to a significant degree the nutritive value or the concentration of toxic constituents therein." (June 25, 1971, 36 F. R. 12093). This regulation will result in greater testing of new varieties before release, an expansion of effort that will benefit all, but that will also materially increase the costs and time of variety development.

MARKETING CONDITIONS

For convenience, vegetables are sold by weight or by volume. In either case the variety that can be placed on the grocer's shelf at the most competitive price will be preferred by management. Unless a variety with unique flavor or other assets has, in addition, a unique, identifiable feature, e.g., shape and russetting of Russett Burbank potato, or shape, color, and aroma of the Red Delicious apple, the housewife will not pay a premium. Accordingly, nearly all varieties remain competitive on the basis of their cost per unit of weight or volume, not on their intrinsic dietary or culinary value. Experimental attempts to sell items graded on the basis of their dietary or culinary quality have never succeeded where a price differential was associated with the quality differences. Market management cannot afford to wait for public acceptance of these attempts, however laudable.

---

* The term antimetabolite is used here in the broad sense to refer to materials that depress growth of test animals maintained on otherwise complete diets.

GENETIC VULNERABILITY BY CROPS

THE CABBAGE FAMILY

The cultivated species of this family are cross pollinating, a fact that implies great genetic diversity within varieties. During the past 20 years, Japanese and American plant breeders have successfully produced hybrids, using self-incompatibility as a means of facilitating controlled cross pollinations between inbred parents. Hybrids, uniform in head size, shape, and color are also uniform in maturity, which makes machine harvesting possible. The extreme uniformity of the hybrid does create greater potential crop vulnerability than was true of the open-pollinated varieties. The inbreeding of cabbage, broccoli, and Brussels sprouts tends to isolate the existing germ plasm into highly uniform genetic subpopulations and tends to eliminate much of the genetic variability.

The capacity to resynthesize a vast array of useful hybrids from a relatively small stable of inbreds does provide commercial flexibility to meet the needs of industry. With the inherent loss in germ plasm through inbreeding, it is unfortunate that public support for breeding of all the cruciferous vegetable crops in 1969 was a meager $71,000. Only 1.2 scientific man-years were committed to systematic genetic change on behalf of the public.

Resistance to cabbage yellows, a soil-borne disease, is controlled by a single dominant gene, which is carried by nearly all United States varieties. Japanese varieites, however, are susceptible. There is also a multigenic type of resistance in the Wisconsin Hollander variety, but no one has incorporated multigenic resistance into other open-pollinated varieties or hybrids. This is one of the few examples of a backup system of disease resistance that exists, but unfortunately it has not yet been incorporated into usable new materials. This reflects the highly competitive nature of the seed industry and the inability of public institutions to carry out needed basic programs. Fortunately, the single dominant gene for resistance has been adequate thus far to avoid an epidemic.

Downy mildew, caused by *Peronospora parasitica*, is serious on broccoli and cauliflower in the Pacific Northwest and western New York as well as on cabbage in Florida. A single gene for resistance was discovered in New York, and almost immediately a new race of the disease organism showed up.

Resistance to club root in cabbage is limited to specific races of the organism, of which several biotypes are known.

Resistance to powdery mildew and *Rhizoctonia* head rot, as they occur in Wisconsin at least, has been shown in each case to be controlled by a single dominant gene.

Resistance to internal tip burn in cabbage is quantitatively inherited, and the resistance isolated by mass selection appears to be stable, and highly recessive.

Resistance to cabbage mosaic is determined by high dominant multigenic inheritance that works well.

Black speck in cabbage appears to be nonparasitic. When the first fresh-market hybrids were grown commercially in Florida, black speck became serious; however, these newly developed and apparently highly desirable hybrids might have been sensitive to the copper used in the spray programs in that area.

When a number of distinct resistances have been accumulated in one or more lines, there is often opportunity to develop multiresistant varieties. The program of cabbage breeding for disease resistance begun about 1915 has led to the recent release of just such a hybrid variety, Sanibel. Through inbreeding, a self-incompatible line breeding true for yellows resistance has been developed; it is highly resistant to mosaic and to internal tipburn but susceptible to powdery mildew and *Rhizoctonia* head rot. This variety has been crossed with an open-pollinated variety (and thus not highly restricted in germ plasm), Globelle, which is not only very uniform in type but also resistant to yellows, tipburn, powdery mildew, and *Rhizoctonia* head rot, although it is susceptible to mosaic. The hybrid is desirable and very uniform in maturity as well as being resistant to yellows, mosaic, internal tipburn, powdery mildew, and *Rhizoctonia* head rot. The use of an open-pollinated parent tends to broaden rather than narrow the germ plasm base. The introduction of resistance to black rot and black speck is under way.

Such a program as that noted above can hardly be expected to be conducted by industry. Much vegetable breeding like this is possible, but it is restricted by lack of public support.

THE NIGHTSHADE FAMILY

Tomatoes, peppers, and eggplant are self-pollinating. Tomatoes are particularly important as a source of vitamin C in our diet, and all varieties are highly inbred populations with no significant genetic diversity within a variety. Self-pollination in nature eliminated variability in these crops long before man tried his skill. Accordingly, much of the needed variation has been introduced by breeders from wild nightshade relatives. The tomato industry is highly dependent on a small number of genes, but the vulnerability that might be associated with these genes is unknown. Industry dependence in tomato would include:

1. a small number of genes for fruit color;
2. genes associated with adaptation to mechanical harvesting to give concentrated set, easy fruit removal on shaking and the Sp gene for determinant growth;
3. the I gene for immunity to *Fusarium* wilt caused by *Fusarium oxysporum lycopersici*;
4. the Ve gene for resistance to *Verticillium* wilt caused by *Verticillium albo-atrum*;
5. the Sm gene for resistance to *Stemphylium solani,* a disease important only in the southern half of the United States.

In addition, many new varieties grown in areas where nematode damage is serious, e.g., Hawaii and Florida, carry a single dominant gene for resistance to that nematode, *Meloidogyne incognita.* Resistance was derived from *Lycopersicon peruvianum*, a near relative of the cultivated tomato. This gene is of commercial value only if used judiciously with crop rotation; otherwise, the nematode populations soon become unmanageable.

Nearly all greenhouse varieties of tomato carry resistance to leaf mold, caused by *Cladosporium fulvum*, but there are many races of the organism and resistance is specific for each race.

Genetic vulnerability obviously exists in the highly uniform varieties of tomato. For example:

● VF 145, released by the University of California, Davis, in 1960 is now grown on about 150,000 acres in California. The rate of acceptance by industry obviously has its justification, but the degree of utilization of a single variety should cause great concern about genetic vulnerability in that area.

● The I gene for immunity to *Fusarium* wilt was thought to offer satisfactory control. However, Alexander at Ohio State has found a strain of the parasite, limited thus far to Florida, Arkansas, Brazil, and Israel, that is not controlled by the I gene. However, a gene for immunity to the second race has been found and is in use in Florida, where a double resistant variety has already been released.

THE VINE CROPS

The vine crops, or cucurbits, include watermelons, muskmelons, squash, pumpkins, cucumbers, and gourds. All are predominantly cross pollinated by honeybees, and the open-pollinated varieties are very heterogeneous. Hence, these varieties generally are not too vulnerable; on the other hand the cucumber industry has had serious disease problems with which to contend and only through diligent efforts of public breeders has resistance been found and incorporated.

Since about 1955 a dramatic change from open-pollinated varieties to hybrids has occurred. The quick acceptance of hybrids resulted because of their uniformity, and their unique and different pattern of fruit production. The pressure to shift from hand harvesting to machine harvesting was directly correlated to concentration of fruit set and the predictable uniformity of maturity of hybrids. In no major crop has the shift from open-pollinated varieties to hybrids been so rapid. Obviously the entire industry has accepted the potential genetic vulnerability associated with the monoculture of these types.

Cucumber scab or spot rot, caused by *Cladosporium cucumerinum*, has been controlled effectively by a single dominant gene for the past 15 years. However, the fungus is moved readily by wind. If, as in corn, a mutant form of the pathogen arises that is not controlled by this gene, the disease will become rampant again.

Mosaic virus is also a serious disease of cucumbers, and it too is controlled quite well by a single dominant gene.

Downy mildew, caused by *Pseudoperonospora cubensis*, and powdery mildew, caused by *Sphaerotheca fuliginea*, are gradually being controlled in cucumbers by genetic resistance. In melons, resistance to powdery mildew, caused by *Sphaerotheca fuliginea*, is controlled by a recessive gene effective against Race 1 but not against Race 2. Resistance is known to Race 2, but its use by industry awaits appropriate variety development. In the meantime, chemical sprays are still necessary to control the fungus.

Melons are also attacked by two different strains of watermelon mosaic virus. Resistance is known for watermelon mosaic virus 1, but not for watermelon virus 2, the prevailing strain in southwestern United States.

Two interesting vulnerabilities do exist in the vine crops:

1. All cytoplasm in western United States varieties of cantaloupe trace back to the variety Hales Best. Although this individual cytoplasm has no apparent unique manifestation, the widespread culture of a single cytoplasm is a potential danger.

2. Most cultivated cucurbits have bitter plant parts but have genetic systems that inhibit production of the bitter substances (called cucurbitacins) in the fruit. Cucurbitacins are powerful insect attractants for the cucumber beetles *Acalymma vittata*, *Diabrotica undecimpunctata howardi* and *Diabrotica balteata*. Cucurbits with a reduced level of bitter substance in the vegetative parts are resistant to these insects, either under mixed stand conditions where the insects have a choice of food plant, or under pure stands where the insects have no choice.

Quantitative differences in the amounts of cucurbitacins in species of cultivated cucurbits appear to be governed by many genes, but single recessive genes in several species (cucumbers, watermelons, and squash) block biosynthesis of cucurbitacins. This simple genetic system in cucumber governs resistance to cucumber beetles associated with nonbitterness. However, the nonbitter plants are susceptible to the two-spotted spider mites (*Tetranychus telarius*). It appears that cucurbitacins originally evolved as protectants against insect feeding. Certain groups of insects (cucumber beetles) first evolved detoxification mechanisms for the protective substances and behavioral modifications that permitted them to locate the food source on which they alone could feed with impunity. However, the protectants still function against other pests. Since mites tend to develop insecticide resistance at a more rapid rate than beetles, this type of resistance must be utilized with caution.

This general relationship seems to hold for other insect-plant associations and may limit the usefulness of breeding for insect resistance through the nonpreference mechanism.

LEAFY VEGETABLES

Spinach is cross-pollinated. Although genetic vulnerability is therefore unlikely, this vegetable has not adjusted to urbanization. Spinach production as a commercial enterprise is disappearing in industrial areas; ozone and other atmospheric pollutants are often blamed.

Lettuce is self-pollinated and has many of the potential weaknesses discussed for tomatoes. Lettuce growers in California need effective control of downy mildew, caused by *Bremia lactucae*, and common mosaic virus. The variety Calmar developed by the University of California and the USDA resists the prevalent race of downy mildew in the Salinas, California, area. Until one or more of the other races becomes dominant, the uniform culture of this one variety will probably continue. The use of the variety Calmar on such a broad scale plus the high level of resistance in this variety greatly decreased the amount of the fungus in the Salinas Valley.

Recently, resistance to common mosaic virus of lettuce (one recessive gene) has been transferred from a leafy noncommercial type lettuce to a crisp-head type commercial lettuce. If generally accepted, such varieties have the potential of genetic vulnerability. In another development, a wild species, *Lactuca virosa*, was used as a female parent to develop a widely adapted variety Vanguard. It is entirely possible that the foreign cytoplasm of *L. virosa* will be combined generally with the genetic materials from *L. sativus*. This type of imbalance of foreign cytoplasm in commercial varieties has long been recognized as undesirable. The potential of the wide culture of *L. virosa* cytoplasm would be entirely analogous to T cytoplasm in maize.

Many plants produce an antimetabolite called gamma amino butyric acid (GABA), a nonprotein amino acid. It is known to be a physiological regulator in some animal tissues, but little is known of its effects when ingested. GABA has been isolated and identified as a hypotensive agent from lettuce. GABA has also been noted to increase in lettuce plants as a response to ozone and also to ammonia or ammonium nitrate in nitrogen fertilizer.

ONIONS AND CARROTS

Onion is a cross-pollinated crop in which, as in maize, a cytoplasmic-genic type of pollen sterility is used in inbreds to ensure controlled hybrid seed production. Thus far, no problems related to the sterile cytoplasm used commercially have arisen even though it can be traced back to a single bulb.

The uniformity of hybrid onions has been of great commercial value. Uniform maturity in northern long-day bulbing-type onions allows mechanical harvesting. All plants mature at the same time, thereby ensuring uniform storage quality and a dramatic reduction in the amount of storage soft rot, caused by *Erwinia carotovora*, and neck rot, caused by *Botrytis allii*.

Resistance to many diseases of onions has been identified and incorporated into onion inbreds and hybrids. Onion breeders also have available valuable levels of resistance to ozone (a single dominant gene). In no case has the onion industry developed total dependence on genes for disease resistance, but this will come with time.

Carrots, like onions, are cross pollinated. The primary efforts of the 1.5 scientific man years at public institutions are largely devoted to the development of inbreds with cytoplasmic-genic control of pollen sterility and the development of uniform hybrids that are not only bright orange and uniformly colored throughout, but also significantly higher than commercial varieties in *alpha* and *beta* carotene. These two pigments are converted to vitamin A when eaten by humans.

At the present time it is not clear whether genetic vulnerability presents a problem. Onions and carrots, however, share a total assignment in the nation of 2.5 public employees, hardly a comforting number. The intensive inbreeding programs cannot maintain the diversity of germ plasm needed, and the acceptance of hybrid onions has rapidly removed open-pollinated varieties as sources of new germ plasm. The short storage life of onion seed makes the problem even more acute.

SUMMARY

The need for high levels of uniformity in vegetable crops plus the necessary dependence on genetic controls for disease resistance make the vegetable industries of the United States vulnerable to disease epidemics. Dependence upon genetic controls cannot and will not be changed. However, thus far, serious outbreaks have not been common. The successful use of resistance to *Fusarium* wilt in cabbage and scab in cucumbers attests to the stability and justification for employing disease resistance based on single genes. The recent dramatic performance of Calmar head lettuce must also be noted.

The value of commercial first-generation hybrids in cross-pollinated crops has far outweighed the risks that might have been anticipated. Hybrid onion seed supplies have never met the demand. Hybrid cucumbers and cabbages have rapidly replaced standard open-pollinated varieties. The vulnerability of male-sterile cytoplasm in onions and the all-female character in cucumbers required for hybrid seed production remains to be demonstrated.

Still, there is little justification for complacency. The interaction of bitterness/nonbitterness in the cucumber crop to insect resistance and susceptibility, the rapidity with which new races of organisms can appear, the identification of antimetabolites in relatives of lettuce, and the introduction of foreign cytoplasm into commercial varieties all give us cause for concern. The disappearance of old varieties displaced by productive, uniform hybrids may be a serious loss of germ plasm for future breeders.

Breeders in the public institutions are quite aware of the many changes in the vegetable industries as well as the increasing number of problems raised by the consuming public, but they cannot make all the changes needed to adapt vegetables to the new requirements rapidly enough. The problems confronting the breeder of onions and carrots are just as many and just as acute as the problems facing corn breeders; the problems of lettuce and tomato breeders are just as many and just as acute as the problems of the cotton or oat breeder. Yet the manpower that is assigned to this segment of agriculture is extremely limited. The very paucity of this commitment aggravates the vulnerability question.

This chapter has addressed itself primarily to the role of public agencies in vegetable breeding in the United States. Significant efforts have also been carried out by seed companies and processors. The interdependence of these groups and the willingness to exchange information and germ plasm have been unique. Industry cannot, however, replace the public agencies as vehicles to solve the high risk projects that need to be undertaken in breeding if we are to progress.

SUGGESTED READINGS

Walker, J. C. 1941. Disease resistance in vegetables. Bot. Rev. 7: 458–506.
Walker, J. C. 1953. Disease resistance in vegetables. Bot. Rev. 19: 606–644.
Walker, J. C. 1965. Disease resistance in vegetables. Bot. Rev. 31: 331–380.

CHAPTER 15

Cotton

Contents

UPLAND COTTONS	271
History	271
Genetic Base	271
AMERICAN PIMA COTTONS	274
MAJOR BREEDING PROBLEMS AND	
PROGRAMS	274
Fiber Quality	274
Diseases	275
Insect Pests	276
Cytoplasmic Diversity	278
RESOURCES AND REQUIREMENTS	279

According to available (1970) statistics, annual cotton production in the United States amounts to a little over 10 million bales, harvested from 11 million acres. Of the total production, less than 100,000 bales are contributed by American Pima varieties (*Gossypium barbadense*), the great majority being produced from varieties of Upland (*Gossypium hirsutum*) cottons. Of the American Pima production, 90 percent comes from Texas and Arizona, the remainder from California and New Mexico. Several varieties of Upland cottons are grown throughout the Cotton Belt from the Eastern Seaboard to California.

Both *G. hirsutum* and *G. barbadense* are native to the New World tropics and subtropics. Thus cottons, grown as annuals in this country, are basically perennial trees or shrubs and non-frost-tolerant; their flowering is restricted to the winter (short-day) months. In order for these species to become established in temperate latitudes, they had to be converted from the perennial to the annual habit and also to develop the ability to flower and set a crop during the summer (long-day) months. From records we know that this transition had been accomplished by French and British colonists in the southeast United States by mid-eighteenth century or earlier. The source

270

of the early annual varieties is obscure, but it is certain that many attempted introductions failed, and that those that succeeded and became the ancestors of our present-day annual varieties had survived an intense process of natural selection. This was the first "bottle-neck" that must have eliminated a considerable amount of genetic variability from United States cottons.

UPLAND COTTONS

HISTORY

At or about the time of the Mexican War (1846–1848) big-boll varieties of cotton were introduced from Mexico through Texas and their germ plasm was combined with the types already established in the interior upland regions of Virginia, the Carolinas, Georgia, and Florida. The resulting conglomerate provided the primary gene pool for modern Upland varieties. A tremendous number of varieties was established by individual farmers and seed merchants during the latter part of the nineteenth century.

Around 1900 the bollweevil (*Anthonomus grandis*) entered the United States from Mexico and spread rapidly over the central and eastern Cotton Belts. Lack of adequate control measures led to elimination of most of the established varieties. There was a strong natural selection for types with determinate growth and early fruiting habit—capable of setting a crop in advance of the seasonal build-up of weevil populations. Thus, the Upland cottons and germ plasm were attenuated even more.

GENETIC BASE

Today the leading strains of three varieties of Upland cottons (Deltapine, Stoneville, and Acala) account for more than 50 percent of the acreage planted.* These are:

1. *Deltapine 16* (26 percent)—a widely distributed variety, bred and developed in the Mississippi Delta, and well adapted to the "river-

* Data from USDA Consumer and Marketing Service, Cotton Div., Memphis, Tenn., July 1970.

bottom" soils not only of the Mississippi River, but other such areas in Arkansas, Louisiana, and Texas. It is also grown in considerable quantity in certain areas in the Southeast, in the Coastal Bend of Texas, and under desert conditions, in southern Arizona and the Imperial Valley of California.

2. *Stoneville 213* (16 percent)—also bred in the Mississippi Delta and adapted to alluvial soils along the Mississippi River and other streams in Arkansas, Louisiana, and Texas.

3. *Acala SJ-1* (11 percent)—bred in the San Joaquin Valley of California and grown mainly in that valley.

Of other cotton varieties that are grown in considerable acreages, but that individually, make up less than 10 percent of the United States acreage, the following should be mentioned:

1. *Lankart 57* (8 percent)—distribution restricted mainly to Texas and Oklahoma.

2. *Coker 201* (4 percent)—grown mainly in the Carolinas and in Georgia.

3. *Paymaster 111* (3 percent)—restricted mainly to the high and rolling plains of Texas and to western Oklahoma.

Of these varieties, the ancestors of Deltapine, Stoneville, and Coker trace to varieties that survived the bollweevil of the early 1900's. They have generally similar fiber properties and relatively few differences in plant habit. Lankart, through sharing a common ancestry with Stoneville, has developed a distinct plant habit and range of adaptation. Paymaster was developed through hybridization and selection from "Kekchi" (a primitive race from Guatemala) and "Macha" (a short-staple type with a stormproof boll). Lankart and Paymaster have short- to medium-staple fibers, adapted to harvesting with the mechanical stripper, and are grown in the prairie and plains areas of Texas and Oklahoma. The Acala cottons, grown mainly under irrigation in the Southwest, have a different ancestry. They trace to a few plants introduced directly from Acala, Mexico, by Cook and Doyle in 1907. They were not subjected to the bollweevil epidemic of the early 1900's and did not initially share the "eastern" germ plasm common to the other varieties, though in recent years they have become infused with it.

Although history and the limited number of varieties grown on a large scale today suggest that Upland germ plasm is very narrowly based, there are mitigating circumstances. An Upland cotton "variety" is neither a pure line nor a primary mixture of pure lines. For the major part of their history, varieties have been developed by single-plant or progeny-row selection (supplemented by intervarietal crossing), followed by open pollination, bulking, and mass multiplication. The effective breeding system, therefore, depends on the relative availability and activity of major pollinating insects. The amount of outcrossing realized may vary from 5 percent in the Southwest to more than 50 percent in Tennessee and the Carolinas. Under the higher levels of outcrossing a considerable amount of genetic variability can be conserved and new recombinations generated. It is still possible by selection, or intervarietal crossing, to isolate strains differing, for example, in wilt tolerance or other desirable properties. The fact that Upland cottons may be restricted to one "variety" in a particular state or region does not imply, in a strict sense, the establishment of a monoculture. It is possible within a given "variety" to develop different strains adapted to particular areas. Furthermore, no single nuclear or cytoplasmic factor has been systematically introduced into a large number of cotton varieties, as has been the case for reduced plant stature in wheat and for male sterility in corn.

On the other hand, it is becoming evident that substantial resistance to the major insect pests and to some of the major diseases is unlikely to be found within the limits of present-day Upland varieties. Also the possibility of improving fiber properties from the same source material is extremely limited. Over the past 30 years there has been a growing interest in the use of exotic material in experimental breeding stocks; with a few significant exceptions, much of this material has not yet been incorporated successfully in practical breeding programs. It is probable that the progress that has been made over these years owes much to factors other than an unlimited supply of genetic variability. Among these are improvements in breeding methodology, better and increased use of fertilizers and insecticides, and the long-term trend from small-scale farming in marginal areas of the Southeast to large-scale monocultures of cotton under near-optimum growth conditions in the Mississippi Delta and California. Improvement in fiber quality certainly owes much to the remarkable development of instrumentation and methods of rapidly measuring both standard and novel components of fiber properties. These precision tools were not available to the breeder 20 years ago.

AMERICAN PIMA COTTONS

The earliest varieties of *G. barbadense* grown in this country were the Sea Island cottons. However, they never became established beyond the Eastern Seaboard and were abandoned in the 1940's. Until the end of the eighteenth century the cottons grown in Egypt were perennial types. Annual types of *barbadense* first appeared in Egypt at the beginning of the nineteenth century; their origin is obscure. Their long staple and fiber strength aroused interest in this country, and introductions were made around 1900–1908. From these early introductions were developed the first American-Egyptian (Pima) cottons. They were adapted to growth under irrigated conditions in the Southwest, but compared with Upland cottons their yield was low. The Pima germ plasm base was then enlarged successfully by hybridization with other varieties of *barbadense* (Sea Island and Peruvian Tanguis) and by an infusion of *hirsutum* germ plasm (Stoneville). The release of Pima S-1 in 1951 was a breakthrough in varietal improvement, and the variety provided a new line of ancestry for the subsequent series of American Pima varieties. Currently the germ plasm base is broad enough to provide steady varietal improvement.

MAJOR BREEDING PROBLEMS AND PROGRAMS

FIBER QUALITY

Major improvements in the fiber quality (particularly fiber strength) of Upland cottons have come through the use of exotic source material. The first of these was Hopi—a primitive form of *hirsutum* grown by the Hopi Indians of New Mexico. Hopi germ plasm was introduced into Acala breeding programs in the 1950's and has since been incorporated into other varietal improvement programs.

A second, and potentially more fundamental improvement has come from the use of the so-called Triple Hybrid that was derived from crossing three widely different species (*arboreum, thurberi,* and *hirsutum*) 30 years ago. Within the last 10 years at least 7 new varieties or breeders' strains incorporating Triple Hybrid germ plasm have been released by experiment stations in 6 states. The history of this material has an important bearing on the general problems of conserving germ plasm and "correcting" genetic vulnerability to epidemics:

1. The original hybrid was not made for the purpose of improving fiber strength. Its potentialities only became apparent "after the fact."

2. Probably no more than six years' selection and breeding were sufficient to establish, on a preponderately Upland background, lines with improved fiber properties. It required 15 more years of intensive breeding to combine these properties with other agronomically acceptable characteristics. These facts emphasize the need to preserve a wide range of material in germ plasm collections, quite apart from their *apparent* potentialities in breeding programs. They also illustrate the difficulty of providing back-up material on an emergency basis when a current variety becomes vulnerable. When exotic material plays a major role as a breeding source, the long time span between the discovery of a desirable character and its incorporation into an economically acceptable strain creates a replacement problem.

DISEASES

Of the four major diseases of cotton in the United States three are soil-borne: *Verticillium* wilt, *Fusarium* wilt, and Texas root-rot. Although all three of these diseases may be locally destructive, none has a beltwide distribution. Thus, from the point of view of genetic vulnerability, it is likely that the emergence of new and more virulent strains of these pathogens could be locally contained—by quarantine regulations or by redistribution of acreage. *Verticillium* wilt may be an exception to this generalization.

Verticillium wilt, caused by *V. albo-atrum*, is the most troublesome cotton disease west of the Mississippi. It is an increasing problem in the San Joaquin Valley of California, in the high plains region of Texas, and in the Mississippi Delta. A more virulent strain of the parasite appears to be spreading in California. There is no known source of resistance; certain strains of *barbadense* exhibit degrees of tolerance. These strains are being used to transfer tolerance into Upland varieties, but since this involves interspecific hybridization, it is a long-term project.

Fusarium wilt caused by *F. oxysporum* f. *vasinfectum*, in the past was a major problem in the Southeast, where it occurs in combination with nematode infestation. Acceptable levels of resistance have been developed in Upland varieties grown in this region. However, since *Fusarium* wilt tends to be a problem of poorer soils, the shrinking cotton acreage in the Southeast and the restriction of the crop to the better soils has reduced the importance of the disease. It is probable that the varieties grown in the Southeast today are less resistant than the older varieties they replaced.

Texas root-rot, caused by *Phymatotrichum omnivorum*, is a major problem only in the black lands of Texas. There is no known source of resistance.

Bacterial leaf blight, caused by *Xanthomonas malvacearum,* creates a different type of problem in two respects: (1) Its potentialities for rapid dissemination are greater; and (2) sources of resistance are known and understood genetically. The breeding program has much in common with that of rust resistance in wheat: the isolation of specific genes resistant to new strains of the pathogen and their incorporation in existing varieties. The main source of resistant (B series) genes so far have been resistant strains of *barbadense.*

INSECT PESTS

Insect control is probably the major problem for cotton growers in all areas except the Far West. The bollweevil (*Anthonomus grandis,* Figure 1), has been established since the beginning of the nineteenth century and the cotton bollworm (*Heliothis zea*) for 150 years. Damaging outbreaks of the pink bollworm (*Pectinophora gossypiella*) have occurred in the states bordering Mexico, and it has become a continuing problem in the Imperial Valley of California, Arizona, and southern Texas. More recently the tobacco budworm (*Heliothis*

FIGURE 1   Adult bollweevil (*Anthonomus grandis*) feeding on a cotton boll. (Photo courtesy S. G. Stephens, North Carolina State Univ.)

virescens) has become the major problem for cotton growers in the Rio Grande Valley. There are no appreciable levels of genetic resistance or tolerance to these insects in any varieties grown in the United States today.

Over the past 25 years insect pests have been contained, but barely controlled, by massive doses of insecticides. Clearly this system cannot be maintained on the general grounds of environmental degradation and specifically in light of the consequences of building up resistance to insecticides. Pilot programs have been started that involve intensive screening of a wide variety of germ plasm, particularly exotic material (wild species, primitive races). Laboratory tests have uncovered a variety of potential sources of resistance to the bollworm. Among these are the "X-factors" that slow down the rate of larval growth in the flower bud or in extracts made from them; high gossypol content in the boll, which is a feeding deterrent; hairless stem and foliage, which reduces egg-laying sites for *Heliothis*; absence of nectaries, which reduces the food sources for the *Heliothis* moth; and attenuated bracteoles surrounding the boll, which reduce oviposition by the bollweevil. All the foregoing characters with the exception of the X-factors are known to be inherited fairly simply, and there should be no insurmountable difficulty in combining them. But it is not yet clear whether the laboratory and small field tests will be confirmed in large-scale culture. Moreover, the nature of the source material (wild and primitive races of *hirsutum*, wild species like *G. armourianum* and *G. sandvicense*, and "developmentally abnormal" mutants of Upland cotton) makes it evident that incorporation into commercially acceptable varieties will require a long-term effort. Meantime "multipronged" programs of bollweevil control are being developed that will attempt to combine the more readily available sources of genetic deterrents with diapause control, strategic rather than massive use of insecticides, and the introduction of sterilized male weevils.

To some extent the question of genetic vulnerability to insect pests in cotton is academic. Cotton is already vulnerable and has been for a long period of time. If the current insect resistance programs are eventually successful, the crop will become less vulnerable than at any time during its past history in this country. Further, the program has a built-in bonus—it cannot fail to enlarge the germ plasm base of Upland varieties and breeding stocks. This should be beneficial for purposes other than those for which the program is designed.

From the long-term point of view, two complicating factors may be anticipated. First, *Heliothis zea*, the cotton bollworm, is also the corn earworm and soybean "podworm." Raising the level of resistance in the cotton plant may have a depressing effect on the total *Heliothis* populations, or it may intensify their attack on corn and soybeans. Presumably a multipronged program of insect control will take into account these broader implications. Second, in recent years there has been considerable interest in the economic development of glandless cottons. These lack gossypol, the toxic organic compound that is present in all normally glanded cottons, and that restricts the use of cottonseed meal as a feed to certain classes of livestock (including humans). The use of cottonseed as a source of protein in human food is more than an academic possibility. Cottonseed meal is the major component of "Incaparina," the protein dietary supplement that has proved to be widely acceptable in underdeveloped countries for the alleviation of the protein deficiency disease, *kwashiorkor*. In the United States there are potential markets for glandless cotton seeds in the flour, candy, and soft-drink trades. A high level of gossypol seems to be a deterrent to insect attack, and small experimental plots of gossypol-free cottons are often attacked by numerous pests that do not normally bother cotton. It does not necessarily follow that the increased vulnerability shown in small plots would be exhibited also in large-scale plantings. Nor is it likely that there will be a rapid switch from glanded to glandless cotton cultivation. Nevertheless, the potential for increased vulnerability is apparent.

CYTOPLASMIC DIVERSITY

Taken at face value, cytoplasmic diversity is not an appropriate topic for discussion under the "major breeding problems" of cotton. It may not be a problem; it has certainly been ignored. However, the discovery of cytoplasmic vulnerability in corn makes it desirable to assess the situation in other crops.

In cotton, cytoplasmic effects have only been studied seriously in connection with male sterility. The results appear comparable to those found in other plants; cytoplasm from a widely different species produces male sterility that can, in some cases, be "restored" by a gene from the same source. Because cotton is insect-pollinated, the practical use of male sterility in breeding is not so obvious as in wind-pollinated crops (for technical reasons involving alternative nectar sites and the foraging behavior of the insect).

Cytoplasmic transfers between closely related species, and between different races of the same species, have apparently not been studied. Crosses involving exotic source material are almost universally made using standard Upland strains as female parents. Consequently, the transfer of genetic material is not accompanied by cytoplasmic transfer, and it is probable that all the material now available in breeding stocks is no more diverse cytoplasmically than the varieties in current production. As shown earlier, the history of these varieties makes it likely that no more than three different cytoplasms—eastern, high plains, and western—may be available in Upland cottons. Further, since these have been intercrossed during their recent history, it is entirely possible that no significant cytoplasmic differences exist in Upland cottons as a whole.

In view of the recent indications in corn that a common cytoplasm may be universally susceptible to new races of a parasite, and of our ignorance of the mechanisms, *it would seem a reasonable precaution to make all wide crosses reciprocally whenever possible.* It should be feasible at the cost of a little extra effort to introduce the cytoplasm from several different species and races of *Gossypium* and if necessary to circumvent the initiation of male-sterile effects by the use of appropriate restorer genes. If cytoplasmic effects should prove to be trivial, there would be no disadvantage in having an array of cytoplasms available in the breeding pool. If, on the other hand, cytoplasmic effects should prove to be important, diversity might reduce the risks of future vulnerability. One might speculate also on the possibility of more positive advantages. If a particular cytoplasm may become universally susceptible to one specific parasite, may it not also become universally resistant to another? Again, transferences of various forms of resistance from one species to another are often difficult—successful only when the resistance mechanism is genetically simple. Conventional wisdom attributes failure to the genetic complexity of the resistance mechanism and to its disintegration during the transference process. Is it possible that some forms of genetic resistance cannot be divorced from the cytoplasm in which they normally operate? If so, cytoplasmic transfer could be an important adjunct to the transference of genes.

RESOURCES AND REQUIREMENTS

Because the nature of genetic vulnerability cannot ordinarily be foreseen, the primary defense against it is the maintenance of a representative collection of germ plasm. As breeding problems and objectives change with time, and as the potentialities of any individual plant in a collection may not be revealed until it has been used as a parent, it is important that the collection be based more on diversity than on apparent utility. Ideally a germ plasm collection should serve two functions, like the services available in a reputable bank. First, it should provide a safe deposit, i.e., long-term seed storage in which samples of the material collected can be stored and periodically checked for continued viability. Properly monitored seed storage is an insurance not only against physical loss of material but against the genetic loss that may occur in living plant collections due to repeated subsampling, natural selection, and contamination. Second, it should provide an account from which samples may be repeatedly withdrawn for experimental purposes. Like a checking account it would have to be continually replenished; this requires the establishment of living plant collections.

The United States cotton germ plasm collection is composed of the following (the numbers given are rounded minimal estimates);

1. Type species of *Gossypium*   50
2. Wild and primitive perennial forms of *hirsutum* and *barbadense*   1,000
3. Varieties of Upland cotton (including obsolete forms and foreign accessions)   1,150
4. Annual forms of *barbadense* (including obsolete forms and foreign accessions)   270

Estimated total   2,470

Samples of all this material are preserved in cold storage at Fort Collins, Colorado. Periodic monitoring has shown that seeds stored as long as 10 years show no appreciable loss of viability. It would not be overly optimistic to suppose that cotton seeds may be preserved in germinable condition for at least 20 years. Responsibility for the maintenance of living collections (the "checking account" function) is shared by several experiment stations. The Texas Experiment Station is responsible for the maintenance of Items 1 and 2 in the list, but duplicates of some of the species and primitive races are maintained elsewhere by individual research workers. Because many of the species and races flower only during the winter months, they have to be maintained on a small scale in greenhouses. The Delta Experiment Station, Stoneville, Mississippi, is responsible for the maintenance of the Upland collection (Item 3) and for supplying seed samples to qualified workers. The United States Cotton Station, Tempe, Arizona, provides a similar service for the annual *barbadense* collection (Item 4).

The problem of maintaining the collection of wild and primitive forms of the cultivated species (Item 2 in the list) has not been solved satisfactorily. Unfortunately, it is this part of the collection that is currently in greatest demand for "resistance screening" purposes. Most of the material consists of small trees or large shrubs, which flower and fruit during the winter months. It is impossible to maintain representative samples in greenhouses—even by pooling the greenhouse resources of several experiment stations and by making partial plantings in successive years. The stopgap solution to this problem that has thus far been employed is the establishment of temporary plantings in tropical locations, in particular at a winter seed-increase nursery in Iguala, Mexico. The nursery has satisfactory working facilities, efficient supervision, and an adequate labor force. However, the station is organized primarily for the advancement of experimental progenies and breeders' strains of *annual* cottons during the off-season. It requires the establishment of an annual row-crop system under conditions of good soil fertility, rigid insect control, and a terminal closed season during the summer months. This is scarcely compatible with the establishment of a perennial planting, which should remain *in situ* for two to three years, needs to be spaced and managed as an orchard crop, and requires a lower fertility level.

Under these circumstances there is a natural tendency to limit tropical plantings to times of emergency, i.e., to replenish seed stocks that are running out, or to "rejuvenate" those that are approaching a terminal stage in cold storage. But stocks that for long periods of time exist only as stored seeds are not available for experimental study. It is unfortunate that a younger generation of breeders may grow up with little first-hand knowledge of the tools available to them.

The establishment of a *permanent tropical station for the maintenance and experimental study of the tropical relatives of major crop plants* would seem to offer a satisfactory solution. In the long run such a common facility might be easier to operate than private arrangements made for each individual crop.

PART III

THE CHALLENGES OF GENETIC VULNERABILITY

CHAPTER 16

The Challenges
of Genetic
Vulnerability

Contents

HOW UNIFORM AND HOW VULNERABLE ARE THE CROPS?	286
Consumers and Processors Demand Uniformity	288
GENETIC NATURE OF THE UNIFORMITY	289
Government Encourages Uniformity	291
CHALLENGES TO THE SCIENTISTS	291
Public Support for Research in Plant Breeding	292
Vigilance	294
Tunnel Vision	295
Exotic Pests	295
Back-up Capability–Diversity of Genes	295
Sources of Genes for Diversity	296
Wild Types and Varieties: Local Varieties; Spontaneous Mutations; Induced Mutations	
Wise Use of Resistance	297
Preserving Genes for Diversity	297
CHALLENGES TO THE NATION	298
A Watchdog System	298
Overseas Laboratories; Offshore Laboratories; Quarantine Services	
Agricultural Research Talent	299
A National Monitoring Committee	299
Germ Plasm Resources	299
Plant Introduction; Seed Storage Facilities	
Variety Development	302
Collections of Parasites	303
Mitigation of Loss	303

285

Two points are clear: (a) vulnerability stems from genetic uniformity; and (b) some American crops are on this basis highly vulnerable. This disturbing uniformity is not due to chance alone. The forces that produced it are powerful and they are varied. They pose a severe dilemma for the sciences that society holds responsible for its agriculture. How can society have the uniformity it demands without the hazards of epidemics to the crops that an expanding population must have?

The severity of the dilemma will stimulate some intensive thought by the scientists and policy makers concerned. There will be shifts in philosophy, shifts in the allocation of present resources, and hopefully the allocation of additional resources.

HOW UNIFORM AND HOW VULNERABLE ARE THE CROPS?

The central question is how uniform and vulnerable are American crops? This is best answered by Table 1, particularly in the last column, which shows the percentage of the acreage of each crop that is planted to a limited number of varieties. For example, 96 percent of the pea crop is planted to only two pea types and 95 percent of the

286

TABLE 1   Acreage and Farm Value of Major United States Crops and Extent to which Small Numbers of Varieties Dominate Crop Acreage (1969 figures)

Crop	Acreage (millions)	Value (millions of dollars)	Total Varieties	Major Varieties	Acreage (percent)
Bean, dry	1.4	143	25	2	60
Bean, snap	0.3	99	70	3	76
Cotton	11.2	1,200	50	3	53
Corn^a	66.3	5,200	197^b	6	71^c
Millet	2.0	?		3	100
Peanut	1.4	312	15	9	95
Peas	0.4	80	50	2	96
Potato	1.4	616	82	4	72
Rice	1.8	449	14	4	65
Sorghum	16.8	795	?	?	?
Soybean	42.4	2,500	62	6	56
Sugar beet	1.4	367	16	2	42
Sweet potato	0.13	63	48	1	69
Wheat	44.3	1,800	269	9	50

^{a} Corn includes seeds, forage, and silage.
^{b} released public inbreds only.
^{c} from Table *3*, Chapter 8

peanut crop to nine varieties of peanut. The figures go as low as 25 percent for two varieties of wheat and 42 percent for two sugar beet hybrids. The data in Table 1 have important implications for the nation, the farmer, and the scientist.

In a certain sense the use of pesticides on crops also reflects genetic vulnerability. Pesticides are used on crops that have a high per acre value and for which pest resistance is inadequate. Farmers naturally prefer resistant crops, but they must use pesticides when the breeders are unable to provide them with resistant varieties. From an epidemiological standpoint the pesticide is used to provide an effect on the pest comparable to genetic resistance.

The extent of this reliance on pesticides is illustrated in Table 2, which shows costs of using them for 1966. The relative difference in use of fungicides and insecticides reflected in Table 2 may be attributed to a number of factors:

1. More effort has been expended on resistance to disease than to insects due to the greater difficulties in establishing uniform insect infestations.

2. There is greater availability of germ plasm conveying resistance to disease.

TABLE 2   United States Expenditures on Fungicides and Insecticides, by Crop (1966)

	Fungicides				Insecticides			
Crop	Acres treated (1,000)	Cost/ Acre	Total ($1,000)	Percent of Farm Value	Acres treated (1,000)	Cost/ Acre	Total ($1,000)	Percent of Farm Value
Corn	0^a	0	38	0	21,804	1.81	39,589	.8
Wheat	0.	0	0	0	1,090	.78	850	<.1
Sorghum	0	0	0	0	329	3.29	1,082	1.6
Rice	0	0	0	0	198	1.63	322	.1
Potato	359	10.47	3,758	.6	1,332	5.56	7,405	1.2
Sugar beet	122	9.43	1,150	.4	147	2.15	316	.1
Vegetables	738	5.35	4,081	.2	2,066	8.49	17,540	1.1
Soybeans	0	0	54	<.1	1,496	1.83	2,737	.1
Cotton	207	7.60	1,573	.1	5,588	10.30	57,556	4.6
Totals			68,460				131,629	

Source: Austin S. Fox, Economics Research Service, USDA.

^{a} Less than .5 percent of acres grown.

3. Cost per acre of treatment in relation to the value of the crop must be considered.

There has recently been, however, an increased emphasis on breeding for insect resistance.

Breeding for insect resistance and the development of other methods of control such as the use of hyperparasites (species parasitic on other parasites) take considerable time and effort. If the current trend to reduce drastically the use of insecticides continues there will be a vulnerable period before these other kinds of control are fully developed.

The striking uniformity among American crops is no accident. Strong forces have been exerted by the market and by farmers, to which plant breeders, plant pathologists, entomologists, and others have responded as well as they could in deriving the best compromises obtainable. That the nation has experienced so few epidemics provides convincing evidence that their efforts have been remarkably effective.

CONSUMERS AND PROCESSORS DEMAND UNIFORMITY

In America it is axiomatic that the marketplace sets the priorities; "the customer is always right." For years no poultryman in his right mind would try to sell white eggs in Boston or brown eggs in New York. Similarly, no supermarket would display deep-eyed, expensive, but good quality Irish Cobbler potatoes when it can get handsome, shiny skinned, inexpensive, but soapy Red La Soda potatoes—housewives prefer the smooth skin. The same standard applies even to the pearls produced by Japanese oyster farmers; the market demands a smooth uniform pearl, and others are contemptuously called "baroques." The corn breeder must produce Dent corns in Iowa, but Flint corns in India. The lettuce breeder must produce a Great Lakes heading type, the pea breeder an Alaska or Perfection type, the snap bean breeder a Tendercrop or Blue Lake type, and so on.

Clearly the market wants uniformity. If one breeder or one farmer fails to provide it, the market will turn to another that will. The irony is that if the uniformity encourages an epidemic, the scientist, not the market, tends to receive the blame.

The market is just as insistent on cost as on uniformity. The market wants to pay the lowest price. If one farmer cannot sell it for the price, another will.

In the sense that the market demands uniformity and low costs, it transfers to the farmer the responsibility for demanding uniformity in turn. In order to meet the cost-price squeeze, the farmer first seeks to raise his efficiency per acre and above all per man hour.

Demands for efficiency are really demands for uniformity in a different guise. The farmer must have high-yielding varieties. Because the low-yielding members of the plant population have been eliminated, this too means uniformity. The farmer must substitute machines for men, but machines can't think, again varieties must be uniform.

Seeds are sown by machine. These too must be uniform or they move unevenly and inefficiently through the planter. The seeds must germinate and grow simultaneously, or they leave space for weeds to grow in the row where the cultivating machine cannot go.

Crops must be uniform for harvesting. Tomatoes, peas, and potatoes must ripen at the same time if they are to be machine harvested, because the machine cannot distinguish between a green tomato and a ripe one.

And so it goes, uniformity—always uniformity.

GENETIC NATURE OF THE UNIFORMITY

Genetic uniformity can take many forms. In the case of vegetatively propagated plants, each variety is uniform for all genes except as mutations occur. If one plant is susceptible, they all are. Apples, for example, are vegetatively propagated; without protective chemicals most apple varieties would be subject to an epidemic of apple scab or apple maggot every year.

Such self-pollinated plants as wheat or bean are similar to vegetatively propagated plants in that varieties are much more uniform than outbreeders. This was the case of the Victoria oat that succumbed to the *Helminthosporium* epidemic, the Marquis wheat that went down before a new race of rust, and the Refugee bean that had to be abandoned because of the common mosaic disease.

Corn is an open-pollinated plant and without man's intervention would be genetically diverse. Breeders reduce it to pure lines, however, by inbreeding—the equivalent of self-pollination.

Uniformity need not rest on a single variety, however. It can rest on a single gene or a single cytoplasm, as in the instance of the Texas male-sterile cytoplasm. Uniformity of this kind has now been introduced into commercial varieties of sorghum, millet, sugar beet, and onion, and it could well become important in wheat. Cytoplasmic uniformity is also found in cotton and cantaloupe.

The important question therefore arises: Does cytoplasmic male sterility aggravate disease in any of these? The answer is, "yes." The male-sterile wheats grown thus far in research plots in the United States are especially susceptible to ergot. In India male-sterile sorghum also has shown appreciable susceptibility to ergot. The problem in both wheat and sorghum seems to stem from the fact that the female plants have open florets as a result of inadequate pollination and is confined to seed production fields. These situations must be closely watched.

A given technological advance in crop production often rests on small numbers of genes. Prominent examples are the dwarf wheats and the single-gene dwarfism in the rice varieties that comprise much of the base for the "Green Revolution." Other examples are the monogerm sugar beets, the determinate gene of tomato, and the stringless gene in beans.

If one of these genes is incorporated into many varieties, the crop becomes correspondingly uniform for that gene. If a parasite with a preference for the characters controlled by that gene were then to come along, the stage would be set for an epidemic, which is precisely what happened when the determinate tomato was introduced in the early 1940's. Varieties with that gene were very susceptible to *Alternaria solani* until new genes for resistance could be incorporated.

Wheats with the dwarf gene are showing susceptibility to the fungus *Alternaria triticina* in the Punjab in India. Dwarf wheats in Mexico are also unusually susceptible to the leaf blotch caused by *Rhyncosporium* sp. This is not to say, of course, that they are uniform in other characters—indeed, they certainly are not. But many wheat varieties do contain the dwarf gene, just as many corn varieties contained Texas cytoplasm. In 1970 the corn varieties in use were diverse for many characters but they were uniform for the T cytoplasm character. For a major epidemic to take place the crop need be uniform for only one character, *provided only that the character in question is favorable to the disease.*

Of course, other genes can influence the single gene or the single cytoplasm, as is evidenced in corn. If they exert enough influences, they may thus aid in quelling an epidemic.

While uniformity is a prerequisite to vulnerability, uniformity alone cannot produce epidemics. But when a parasite appears that can exploit the gene or cytoplasm that brings about uniformity, then the uniformity provides a situation in which an epidemic can occur.

GOVERNMENT ENCOURAGES UNIFORMITY

In pursuing uniformity, the farmer is encouraged by state and federal seed agencies, who, for decades, have emphasized the importance of uniformity and purity. The entire seed certification program, for example, has adversely influenced genetic diversity through its efforts in this direction.

The law not infrequently outruns biology; thus federal regulations in Canada designed to promote uniform wheat quality impose stringent requirements on development of new wheat varieties in that country. The indirect effect of this legislation has been to narrow the germ plasm base and increase potential vulnerability of Canadian wheat. A similar effect has developed from the enforcement of a one-variety cotton law in the San Joaquin valley of California and, more generally, from the recently enacted Federal Plant Variety Protection Law (see chapter on wheat, page 119).

CHALLENGES TO THE SCIENTISTS

Eventually the challenge of genetic vulnerability winds up on the desk, and in the greenhouses and field plots, of the scientists.

PUBLIC SUPPORT FOR RESEARCH IN PLANT BREEDING

Support for research falls into several categories. If new varieties have sufficient income potential to an individual firm to justify a breeding program, private capital will support a significant amount of the work. Otherwise, the responsibility is usually left to the public agencies. The relationship between state and federal responsibility for plant breeding is quite analogous to the relationship between industry and public agencies; if the benefits accrue uniquely to an individual state, the state is often willing to support the cost of research. But if not federal support is sought. No one system neatly fits every situation. Industry, state, and federal agencies all tend to share the expense of long-term breeding programs. The proportion of cost borne is the end result of political negotiation. In fact, one of the basic arguments put forth in support of varietal patent protection (Public Law 91-577) was that income derived from royalties or licensing would encourage greater participation by private industry in plant-breeding research.

The priorities set by public agencies for research support must recognize the different needs for a wide spectrum of crops. Vegetable crops must compete with agronomic and forest crops, though the total dollar value of vegetables is usually secondary to crops grown over extensive acreages. (Support for research in plant breeding is listed in Table 3.) It is difficult to determine the minimum number of professional people necessary to insure that the interests of the public are protected. Obviously 65.1 scientific man years committed to corn breeding by public agencies, plus an even larger commitment in the private industries, were inadequate to prevent the losses resulting from southern corn leaf blight. As the number of individuals committed to breeding decreases, the amount of germ plasm actively used by breeders also decreases. Much potential germ plasm is lost for lack of manpower and an inordinate amount of public responsibility must be assumed by a few individuals. The degree to which these few can effectively create, evaluate, release, and store valuable germ plasm is extremely limited.

In this context, the breeder seldom wins. If he appeals his case to the public, he is labelled a public relations man who should spend more time in the laboratory; if genetic vulnerability results directly from his research efforts, he is labelled shortsighted with respect to the limitations of new germ plasm or insensitive to his public responsibility. As a result, the plant breeders in the public sector do what

TABLE 3   Scientific Man Years (SMY) Assigned to Plant Breeding Research (1969–1970)

Commodity Agronomic	Total (SMY)	Commodity Horticultural	Total (SMY)
Corn	65.1	Potato	20.2
Grain sorghum	16.4	Carrots	1.5
Rice	7.5	Tomato	20.8
Wheat	51.2	Bean–pea	20.1
Barley	20.0	Sweet Corn	3.3
Oats	15.2	Cucurbits	10.5
Small grains	23.3^a	Sweet potatoes	3.7
Soybeans	35.9	Crucifers	1.2
Cotton	45.8	Onions	1.0
Tobacco	32.3	Vegetable crops	35.1^b
Alfalfa and other legumes	24.3		
Grasses and other forages	33.0		
Total	370.0	Total	117.4

Source: Analysis of 1969 CRIS reports by H. J. Hodgson, Coop. State Res. Serv., USDA.

^{a} Scientific man years assigned to small grains without specifying crop. This value is in addition to the assigned values for wheat, barley, and oats.
^{b} Includes lettuce and other crops not itemized separately; in addition this value would include commitment to the listed vegetables without specifically designating the programs by crop.

they can within the limits of their support. If they have a source of disease resistance they use it; to develop back-up systems is a luxury they cannot often afford. For the most part, other breeding problems are accorded a higher priority. It is hardly surprising that in 1969 over 50 percent of the plant-breeding work in public institutions was directed to yield (25 percent) and disease and insect resistance (28 percent), while as little as 8 percent was expended on quality factors and 2 percent on physiological and morphological research. These priorities result largely from the demands of industry and the limitation of research resources.

The scientists challenged by the issue of genetic uniformity are not only plant breeders—they are joined by agronomists, biochemists, geneticists, climatologists, economists, entomologists, plant pathologists, and plant physiologists. All, but especially the plant breeder, the plant pathologist, and the entomologist find themselves under a frustrating pressure. Each hopes to serve society well, but society wants uniformity. The scientist provides it, knowing full well that one day his uniform variety may suffer in the face of an epidemic.

Collectively, plant breeders have a most important influence on the amount of genetic uniformity to be found in commercial crop varieties. What, therefore, might the breeder be expected to do to reduce genetic vulnerability of the crops with which he works?

Procedures used by most breeders tend to narrow rather than expand the genetic base of cultivated plants. The most efficient systems of developing improved varieties involve the use of proven, elite germ plasm rather than unadapted "exotic" varieties. Thus, the tendency is to extract new genetic recombinants from crosses of the best germ plasm to be found in widely used commercial varieties and hybrids.

The concerted use of elite sources of germ plasm in no sense takes place because other more diverse sources are unavailable. On the contrary, for most important crops, a vast array of genetically diverse germ plasm is available. Such material, however, is difficult to work with because it is untested and poorly adapted. Consequently, the breeder who is expected to produce positive results within reasonable periods of time naturally chooses to work primarily with the newer, more elite, and best adapted materials. He may feel he cannot afford the time and effort required to screen scores of collections in the hope of finding a few useful genes.

Both public and private scientists have two principal responsibilities in reducing vulnerability. First, they must exercise constant vigilance in detecting new hazards. Second, they must expand and refine their resources for combating disease by providing new parental material. An excess as well as diversity in breeding stocks is the surest measure. In addition the scientist must push forward in his understanding of the basic principles of parasitism. To make maximum use of field resistance he must understand the life history and ecology of the parasite. Lastly, he should strive to elucidate the pathway between genes and the specific attributes of resistance. Only when this last is done can we move to more exact and specific measures of combating epidemics.

VIGILANCE

Freedom from epidemics is purchased at the price of vigilance. Though United States corn breeders tested their Texas cytoplasm against *Helminthosporium maydis* at several locations, they did not test their material against the Philippine fungus. They concluded that in the Philippines, it was the weather, not the parasite, that differed from that in the Corn Belt.

To disperse a uniform variety of whatever type or whatever crop over much of the country is to spread a wide net. If a parasite has mutated anywhere it will be caught. Hence, even a trace of infection, anywhere, should be examined with great care.

TUNNEL VISION

Scientific tunnel vision may have obscured somewhat the significance for corn of the oat epidemic and of the phenomenon of drug resistance. The Texas male-sterile corn was deployed over a wide area, as was DDT against flies, with similar consequences: a new strain of *H. maydis* overcame the Texas strain of corn in the same way that a new strain of house flies overcame DDT.

EXOTIC PESTS

Most, if not all, of the epidemics discussed in this report have been generated by exotic pests. The parasites that attacked French grapes came from North America. Chestnut blight and Dutch elm disease came originally from the Orient, as did the Japanese beetle. The potato blight fungus doubtless came from Central or South America, the boll weevil from Mexico. The Hessian fly moved across the Atlantic with the straw brought along to feed and bed down the horses used by Hessian soldiers in the Revolution.

The list is endless, the message clear. Man generates his own epidemics because he carries his parasites along with him. Not all the possible parasites have yet been brought to the United States. Some are serious abroad on the very crops we grow here at home. The nation has done little to test domestic varieties of crops against the exotic pests not yet arrived. The corn blight is a case in point. American strains of corn could have been taken to the Philippines and tested against the Philippine strain of the fungus as early as 1964.

BACK-UP CAPABILITY—DIVERSITY OF GENES

If uniformity be the crux of genetic vulnerability, then diversity is the best insurance against it. Since the market demands uniformity, the challenge to the breeder is to provide diversity. He must build redundancy into a back-up system.

As for the corn blight epidemic, it must be recognized that breeders, both commercial and noncommercial, did have a highly effective back-up system. When some small defects appeared in the T cytoplasm system, they began to put normal cytoplasm back into hybrid seed production. This program was pushed strenuously during the winter of 1970–1971, when seed from normal cytoplasm was grown in Mexico, Hawaii, Argentina, and other areas. A major proportion of the seed produced in 1971 for the 1972 crop was produced from inbreds with normal cytoplasm.

SOURCES OF GENES FOR DIVERSITY

The breeders have met the challenge to offset uniformity with diversity by searching for new genes and by developing gene pools to preserve those they have.

Wild Types and Varieties   In the case of a few crop plants, the breeder can go back to the geographic area where the crop seems to have originated in a search for useful genes from the wild types occurring there or even from varieties that local farmers grow. This requires trips of exploration, of course.

The primitive varieties and wild types are threatened, however, by the invasion of their homeland by "improved" varieties. These latter have lost many genes in the process of tailoring them to the uniformity required by the sophisticated markets. Nobody can yet assess the magnitude of this threat.

Several bodies have investigated the extent of this genetic erosion: in 1967 the International Biological Program (IBP) sponsored an international discussion in Rome (Frankel and Bennet, 1970), and FAO is also cataloging existing germ plasm collections of major crops throughout the world.

Local Varieties   Scattered over the globe are innumerable varieties adapted to local conditions. These can be collected and serve as a source of miscellaneous genes for disease resistance or other attributes.

Spontaneous Mutations   The Texas strain of *Helminthosporium maydis* dramatically illustrates the truism that organisms mutate. Most crop plant mutants are off-types and farmers rarely save them. Occasionally one does, as in 1922 when a farmer in Enfield, Connecticut observed an ear of corn in his field that had a peculiar look—it was opaque, not translucent. Knowing that Singleton and Jones at New Haven were saving peculiar corn types, he sent it to them. They named it Opaque-2 and held it in their gene pool for 40 years. Mertz and Nelson, at Purdue University, found it to be the source of a gene for high lysine, and therefore of significant potential for improving the nutritive value of corn.

Induced Mutations Man can empirically induce an occasional useful mutation by treating plant parts with ultraviolet radiation, X-rays, gamma rays, or with chemical mutagens.

WISE USE OF RESISTANCE

In the past each new gene for resistance has been pressed into service with little thought of the probable side effects. The analogy as set forth in Chapter 1 between the uses of crop plant resistance in agriculture and the use of antibiotics in human medicine is pertinent here. The appearance of bacteria resistant to penicillin brought about not only a search for new antibiotics but led eventually to restraint in their use. Doctors realized the dangers of squandering antibiotics by using them in trivial ways that allow resistant pathogens to develop. Antibiotics are now available only on prescription.

The plant breeder recognizes the dangers inherent in single-gene resistance. However, until general resistance is found or developed, public pressure requires that he release any type of resistance available that will minimize a current threat even though the longer term hazards remain. There are instances where specific resistance is extremely valuable but endangered by its geographical deployment. Each year oat crown rust sweeps north, blown by prevailing winds from areas where the spores survive the winter. The resistance genes used to control the rust in the South should be different from those used in the North where the disease will finish its course. If they are the same the build-up of virulent races in the South could lead to heavy infestation elsewhere. Achieving wise deployment necessitates widespread agreements that may sometimes transcend local short-term interests. Breeders and pathologists must prescribe wisely to shepherd our resources of resistance and maximize their usefulness.

PRESERVING GENES FOR DIVERSITY

An effective device being used more and more by geneticists and plant breeders to meet the challenge of uniformity is the germ plasm bank or gene pool. The genes preserved there are to be drawn upon when needed; they provide the base for diversity. Gene pools have generally been developed by the breeders themselves and, hence, are widely scattered. This in itself is diversity of a sort, but it has a serious built-in hazard—if a breeder dies or retires, his material may well be lost.

CHALLENGES TO THE NATION

The nation cannot tolerate epidemics of disease or insects in its basic food and fiber crops. Thus it must face up to the threat posed by such epidemics.

A WATCHDOG SYSTEM

One measure to meet the challenge of ignorance is to set up a watch-dog system with several components; some of these are already in existence, others must be established.

Overseas Laboratories   Clearly, a study of exotic pests that are potential threats to our major crops should be considered. In 1966 a plant sciences panel appointed by the National Academy of Sciences emphasized this point, saying

> Some diseases are of little consequence in their homeland abroad, but they can be devastating when they emigrate to the United States. . . . Facilities are, therefore, needed to study diseases and insects abroad, using American varieties and clones, planted where they can show their resistance (or susceptibility) to diseases endemic in Europe, Asia, or Africa.

Specifically, such crops as corn, cotton, sorghum, and millets should be tested against insects and parasites at breeding stations in the tropics where those crops originated.

The varieties should be tested where the exotic pests are, not after they are imported. Yet only a very small amount of work was done in this direction as Public Law 480 funds became available. Most of these funds have now dried up in those countries where our basic food crops originated. As a substitute, such work could be easily established in cooperation with local experiment stations.

Offshore Laboratories   Offshore laboratories to study susceptibility of American crops to exotic pests would be useful. The USDA already operates some facilities of this kind; the Plum Island Animal Disease Laboratory in Long Island Sound was established to investigate the hoof-and-mouth disease of cattle, and the facilities of Mayaguez, Puerto Rico, are being utilized to screen wheat and oats against exotic races of rust. The Committee feels that much more of this sort of testing should be done to assess the vulnerability of our crops to exotic pests.

Quarantine Services   As any tourist can testify, the United States has an effective quarantine service to intercept pests at the boarders. This is, in fact, the very last opportunity to stop them.

AGRICULTURAL RESEARCH TALENT

If an exotic pest is not studied on United States varieties abroad or in an offshore laboratory, and is not intercepted at the borders, it is very likely to be found quickly by an agricultural scientist at a state or federal experiment station. By that time it will be too late to stop it completely, but at least the alert can be sounded and the situation dealt with.

Also important is the amount of effort that can be allocated to research and development designed more clearly to resolve unanswered questions. At present, genetic reserves of some crops are being mined because public commitment of personnel is insufficient to retain reasonable germ plasm resources.

A NATIONAL MONITORING COMMITTEE

Finally the Committee suggests the establishment of a national monitoring committee to keep a watchful eye on the development and production of major crops and to remain alert to potential hazards associated with new or widespread agricultural practices. This would be, in effect, a committee for technological assessment. It could best serve under the auspices of a nationally recognized organization, and should be comprised of scientists from the USDA, the state experiment stations, the universities, industry, and the general public. It should be advisory in nature, but should be free to issue warnings wherever and whenever it feels them justified.

GERM PLASM RESOURCES

Every introduction of variability to combat vulnerability depends on germ plasm resources. These are the raw materials that will be reshaped and fashioned through breeding to produce the crop varieties of the future. They are among our most precious commodities. Their wise and effective management is a vital task.

Many breeders and geneticists maintain their own gene pools, and thus have preserved innumerable genes from being lost. While the United States has no native major germ plasm, its capacity to respond to threats of epidemic pests and diseases rests on the availability of suitable germ plasm. We must face the question whether the national effort is sufficient to meet the need.

Plant Introduction The benefits of introducing new materials from the centers of origin are widely recognized. All plant breeders recognize the need for germ plasm banks in which representative samples of primitive races and their wild relatives, "land races," obsolete varieties, etc., are stored. Other than spontaneous mutations, or those induced experimentally, the genes available in a germ plasm bank are the only primary resource for the plant breeder. Without them he is restricted to making recombinations of the genes already available in his breeding plots. Though the numbers of possible recombinations are enormous they can be lost rapidly as a single variety becomes predominant. Thus, while the need for germ plasm banks is obvious, the means of maintaining them efficiently deserves examination. The maintenance of living collections is extremely expensive and time-consuming, yet not more so than the long-term, cumulative efforts expended by the plant collectors. Furthermore, material collected even as recently as 10 years ago may no longer be available today. According to Frankel we may be approaching a situation where the primitive races of some of our crop plants will exist solely in germ plasm banks—they can no longer be collected in the wild. What is lost now may well be lost forever.

In the United States, the Plant Introduction Service of the USDA has met this challenge well within the resources available to it. (Burgess, 1971). It organizes collecting expeditions in various parts of the world in response to the needs of private and public plant breeders and other plant scientists. The introductions are cataloged, screened for viruses and other parasites, and eventually made available to cooperating scientists through the regional Plant Introduction Stations. Four such centers serve the continental United States; each has special responsibilities, for example, in maintaining and evaluating comprehensive collections or germ plasm banks of one or more crops. These programs have been, and are now, very useful. Indeed, many successful breeding programs began with materials received through the Plant Introduction Service.

Seed Storage Facilities The National Seed Storage Laboratory, Fort Collins, Colorado, maintains collections of seed under storage conditions that minimize deterioration and maintenance costs. These collections include seed of older primitive varieties that are no longer in cultivation.

The combined expenditure by the Agricultural Research Service and the states on the nationally coordinated plant introduction program is of the order of $2.6 million per year.

Long-term seed storage has several advantages. With low temperature storage facilities and adequate monitoring, it may be possible to maintain seed viability for at least 20 years. Consequently it is entirely possible to maintain germ plasm banks of many crops plants as seeds, and avoid the risks and costs of growing thousands of plant collections every few years.

As useful as cold-storage facilities undoubtedly are, they may engender an unwarranted sense of complacency. Once seed samples are carefully selected and stored away, the generation that deposited them is apt to turn to other matters—the responsibility of taking the stocks out of storage, of deciding where to grow them, and what to keep, is left to the next generation. Since stocks maintained solely in cold storage are not available for study and experiment, the next generation may well have had no first-hand experience whatever with the material it finds itself called upon to evaluate. Then, too, problems and breeding techniques change with time, and material that seemed useful in the 1960's may seem 20 years later to have been curiously chosen. There will be a temptation to discard those stocks that are not "relevant" at the later date. Yet what is discarded in the 1980's may well again be relevant in the year 2000. It seems, therefore, that study and evaluation must be a continuous process—both for the well-being of the collections and for the continued awareness of those for whom it is being maintained. One needs not only a "safe deposit" (cold storage) but also a "checking account" (living collection).

The maintenance of a living collection is usually regarded as a routine and time-consuming, yet essential, task. Much of the material maintained there seems to be of little current interest—occasionally, of course, a threatened epidemic or new insight into a disease problem will generate a sporadic interest in screening everything available. A great deal of the material that leaves the bank is discarded; it is regarded as a gift, not as a loan. Surely a more suitable arrangement could be devised. One possibility is the development of germ plasm maintenance centers, in which maintenance is regarded as the primary goal, not as a by-product of breeding activities. In this connection it is worth noting that the plant breeder, entomologist, or plant pathologist tends to regard a germ plasm collection as an inexhaustible source from which he can extract only such experimental material as is of interest for his specific purposes. On the contrary, the experimental taxonomist, crop plant evolutionist, and ethnobotanist consider the entire collection to be necessary experimental material—quite apart from its potential economic value. They need to preserve and study as wide a variety of material as they can manage and to use it continuously. Because all classifications have to be revised as new material is collected and new analytical procedures developed, the experimental taxonomist's work is never done. A germ plasm maintenance center could effectively serve a dual purpose as experimental material for several disciplines and as a reliable and continuing source of germ plasm for the plant breeder. Coupling these two objectives could relieve the plant breeder of a routine chore and provide the taxonomist and others with facilities and experimental material not now available to them.

The problem of organizing germ plasm maintenance centers varies considerably according to the crop. It is probably most straightforward in those plants that are ordinarily self-fertilizing annuals, like wheat and soybeans. The complexity increases when outbreeding plants (like corn) or plants that are perennial and winter- (short-day) flowering are considered. Cotton combines all these disadvantages and a living collection of cotton germ plasm could only be established in a subtropical, frost-free environment. Since many of our crop plants are native to the subtropics, the establishment of a subtropical center should be carefully considered.

VARIETY DEVELOPMENT

It would be foolish to think of the nation's gene pools as warehouses stocked with finished products. They are much more like stores of outdated military surplus equipment awaiting disassembly and reassembly into new, more useful, forms. The items in a gene pool cannot be stored as disassembled parts but only within a living organism whose constituent properties are catalogued. These plant materials are exotic. They are usually unadapted to local conditions and well below the standard requirements for economic use. The development of a new variety that incorporates a new gene for resistance from such a pool takes time. For most crops this is hardly less than 5 years and for some might be longer than 20 years.

The search for genes for resistance and their incorporation in new varieties is a continuous process for many crops. For example, a cereal breeder who releases a new rust resistant variety may well have others in different stages of development that will be replacements for the years ahead. Meeting the challenge of an unanticipated threat exposes the weakness of the means whereby we hold the line during the interim period of varietal development. There is no simple solution. This weakness makes it all the more imperative to avoid genetic vulnerability in every way we can.

COLLECTIONS OF PARASITES

Green plant germ plasm is of first importance, but the nation must also have collections of fungi, bacteria, viruses, nematodes, and insects. The American Type Culture Collection, a nonprofit organization supported by federal and other grants and the fees it collects for services, plays an important role in maintaining cultures of many of these parasites. Such other collections as that of the USDA Northern Regional Research Laboratory at Peoria, Illinois, share this burden. There are no major collections of cultures of plant-destroying insects. While many species can now be reared in the laboratory there is only an incomplete index to existing individual collections in the United States. The increasing interest in crop resistance to insects will surely create a demand for cultures of insect pests in the future.

MITIGATION OF LOSS

Finally, there are economic devices that the nation may use to mitigate the impact of losses from an epidemic. The corn blight epidemic could have been disastrous had we depended more completely on corn, as did the pre-Columbian Indians. The T strain of *Helminthosporium maydis* and a monoculture of T cytoplasm would have been disastrous to their society, even though it was not disastrous to ours. The corn the nation had already in storage dampened the economic impact of the epidemic, a circumstance that supports the principle of the "ever-normal granary." Moreover, unused acreage was available and could be diverted to corn the next year to restock the storage bins.

Crop insurance against the hazards of an epidemic is also available to individual farmers in the United States.

REFERENCES

Burgess, S. [ed.]. 1971. The national program for conservation of crop germ plasm. U.S. Dep. Agric., Agric. Res. Serv., Washington, D.C. 73 pp.

Frankel, O. H., and E. Bennett [ed.]. 1970. Genetic resources in plants—their exploration and conservation. F.A. Davis Co., Philadelphia. 554 pp.

COMMITTEE ON GENETIC VULNERABILITY OF MAJOR CROPS

MEMBERS

JAMES G. HORSFALL, Connecticut Agricultural Experiment Station, New Haven, *Chairman*

GEORGE E. BRANDOW, Pennsylvania State University, University Park

WILLIAM L. BROWN, Pioneer Hi-Bred International, Inc., Des Moines, Ia.

PETER R. DAY, Connecticut Agricultural Experiment Station, New Haven

WARREN H. GABELMAN, University of Wisconsin, Madison

JOHN B. HANSON, University of Illinois, Urbana

*R. F. HOLLAND, DeKalb Agricultural Research, Inc., DeKalb, Ill.

ARTHUR L. HOOKER, University of Illinois, Urbana

PETER R. JENNINGS, The Rockefeller Foundation, Centro Internacional de Agricultura Tropical, Cali, Colombia

V. A. JOHNSON, U.S. Department of Agriculture, University of Nebraska, Lincoln

DON C. PETERS, Oklahoma State University, Stillwater

MARCUS M. RHOADES, Indiana University, Bloomington

GEORGE F. SPRAGUE, U.S. Department of Agriculture, Plant Industry Station, Beltsville, Md.

STANLEY G. STEPHENS, North Carolina State University, Raleigh

JAMES TAMMEN, Pennsylvania State University, University Park

WILLIAM J. ZAUMEYER, U.S. Department of Agriculture, Plant Industry Station, Beltsville, Md.

* Resigned December 13, 1971.

305

SUBCOMMITTEE ON CORN

GEORGE F. SPRAGUE, U.S. Department of Agriculture, Plant Industry Station, Beltsville, Md., *Chairman*
WILLIAM L. BROWN, Pioneer Hi-Bred International, Inc., Des Moines, Ia.
ARTHUR L. HOOKER, University of Illinois, Urbana
MARCUS M. RHOADES, Indiana University, Bloomington

SUBCOMMITTEE ON COTTON

STANLEY G. STEPHENS, North Carolina State University, Raleigh, *Chairman*
CHARLES F. LEWIS, U.S. Department of Agriculture, Plant Industry Station, Beltsville, Md.
PHILIP A. MILLER, North Carolina State University, Raleigh
THOMAS R. RICHMOND, Texas A&M University, College Station

SUBCOMMITTEE ON POTATO AND ROOT CROPS

PETER R. DAY, Connecticut Agricultural Experiment Station, New Haven, *Chairman*
M. E. GALLEGLY, West Virginia University, Morgantown
S. J. PELOQUIN, University of Wisconsin, Madison

SUBCOMMITTEE ON RICE

PETER R. JENNINGS, The Rockefeller Foundation, Centro Internacional de Agricultura Tropical, Cali, Colombia, *Chairman*
C. ROY ADAIR, U.S. Department of Agriculture, Plant Industry Station, Beltsville, Md.
GURDEV S. KHUSH, International Rice Research Institute, Manila, Philippines
NELSON E. JODON, U.S. Department of Agriculture, Rice Experiment Station, Crowley, La.

SUBCOMMITTEE ON SORGHUM AND MILLET

*R. F. HOLLAND, DeKalb Agricultural Research, Inc., DeKalb, Ill., *Chairman*
GLENN W. BURTON, U.S. Department of Agriculture, Georgia Coastal Plain Experiment Station, Tifton
J. ROY QUINBY, Pioneer Sorghum Co., Plainview, Tex.
ORRIN J. WEBSTER, U.S. Department of Agriculture, Mayaguez, Puerto Rico

SUBCOMMITTEE ON SOYBEANS AND EDIBLE LEGUMES

WILLIAM J. ZAUMEYER, U.S. Department of Agriculture, Plant Industry Station, Beltsville, Md., *Chairman*
M. W. ADAMS, Michigan State University, East Lansing

* Resigned December 13, 1971.

BILLY E. CALDWELL, U.S. Department of Agriculture, Plant Industry Station, Beltsville, Md.

RAY O. HAMMONS, U.S. Department of Agriculture, Georgia Coastal Plain Experiment Station, Tifton

M. C. PARKER, Gallatin Valley Seed Company, Twin Falls, Id.

SUBCOMMITTEE ON VEGETABLES

WARREN H. GABELMAN, University of Wisconsin, Madison, *Chairman*

CHARLES M. JONES, Purdue University, Lafayette, Ind.

J. C. WALKER, University of Wisconsin, Madison

THOMAS W. WHITAKER, U.S. Department of Agriculture, U.S. Horticultural Field Station, La Jolla, Calif.

SUBCOMMITTEE ON WHEAT

V. A. JOHNSON, U.S. Department of Agriculture, University of Nebraska, Lincoln, *Chairman*

R. E. ALLAN, U.S. Department of Agriculture, Washington State University, Pullman

ROBERT L. GALLUN, U.S. Department of Agriculture, Purdue University, Lafayette, Ind.

WILLIAM Q. LOEGERING, University of Missouri, Columbia